RADIOTHERAPY TREATMENT PLANNING

RADIOTHERAPY TREATMENT PLANNING
Linear-Quadratic Radiobiology

J. Donald Chapman
Alan E. Nahum

CRC Press
Taylor & Francis Group
Boca Raton   London   New York

CRC Press is an imprint of the
Taylor & Francis Group, an **informa** business

CRC Press
Taylor & Francis Group
6000 Broken Sound Parkway NW, Suite 300
Boca Raton, FL 33487-2742

First issued in paperback 2019

© 2015 by Taylor & Francis Group, LLC
CRC Press is an imprint of Taylor & Francis Group, an Informa business

No claim to original U.S. Government works

ISBN-13: 978-1-4398-6259-9 (hbk)
ISBN-13: 978-0-367-86643-3 (pbk)

Library of Congress Cataloging-in-Publication Data

Chapman, J. Donald, author.
   Radiotherapy treatment planning : linear-quadratic radiobiology / J. Donald Chapman and Alan E. Nahum.
   p. ; cm.
   Includes bibliographical references and index.
   ISBN 978-1-4398-6259-9 (hardcover : alk. paper)
   I. Nahum, Alan E., author. II. Title.
   [DNLM: 1. Neoplasms--radiotherapy. 2. Linear Models. 3. Radiobiology. QZ 269]

   RC271.R3
   616.99'40642--dc23   2014042615

Visit the Taylor & Francis Web site at
http://www.taylorandfrancis.com

and the CRC Press Web site at
http://www.crcpress.com

Dedicated to the medical physicists and dosimetrists who each day plan for and assist with the delivery of radiation oncology prescriptions to patients. Their efforts often go unnoticed but are essential for producing curative cancer treatments.

Contents

Preface

There are currently several textbooks of radiobiology available for teaching the subject to medical physicists, radiation oncology residents and graduate students. Some of these have been updated on a regular basis and attempt to cover the breadth of the sciences (radiation physics, radiation chemistry, cell biology, tissue pathology, physiology and patient management) involved in the process of tumor cell killing, which has become an integral part of current cancer treatments. And, of late, a strong emphasis has been placed on genetic and molecular biology characterizations that might inform about these radiation processes and lead to an improved diagnosis and treatment of cancers. So why another textbook on this subject at this time?

The field of radiation oncology has experienced an immense boost in recent years in its ability to identify gross tumor volumes (GTVs) and planned treatment volumes (PTVs) with advanced techniques of tumor imaging, including computerized tomography (CT), magnetic resonance imaging (MRI) and positron emission tomography (PET) in addition to conventional radiology and ultrasound. Also, novel systems for delivering radiation dose to PTVs have been developed and are operational. These include higher energy linear accelerators for electron and photon beams, systems for tomotherapy, gamma-knives for focal irradiation and systems that deliver beams of neutrons, protons and carbon ions. These improved technologies allow for a much better conformation of the prescription dose to tumor volumes with significant reductions in exposures to normal tissues. The standard practice of ~2 Gy fractions delivered 5 days a week uniformly to PTVs up to the dose that might negatively impact normal tissue function was empirically arrived at over several decades of clinical experience using inferior technologies. Of current interest is whether or not these technology advances will allow for modifications to current radiation prescriptions that prove to be more efficacious, more acceptable to patients (fewer fractions) and possibly less costly. Planning altered fractionation schemes for patient treatment requires reliable quantitative knowledge of the expected radiobiology responses of various tumors and normal tissues.

Over the past 40 years, radiation biophysicists have quantified the killing of tumor cells by radiation in mathematical terms that could be useful for treatment planning. At first the single-hit multitarget (SHMT) formulation was the choice to describe the cell response to radiations administered by different sources in various laboratories. But currently, there is general agreement that the linear-quadratic (LQ) equation defines radiation-induced cell killing more precisely and has now become the standard. The coefficients for single-hit (alpha) and double-hit (beta) inactivation can precisely define the radiation response of cultured human tumor cells when rigorous laboratory standards are employed.

In this book, we will describe tumor cell inactivation from a radiation physics perspective and suggest appropriate LQ parameters for modeling tumor and normal tissue responses. Consequently, the majority of data cited will be from those reports that describe cell killing in terms of the two independent processes of the LQ model. Much of that information has come from our laboratories with the intent of extending it to tumor control probability (TCP) and normal tissue complication probability (NTCP) modeling. The compilation of radiation mechanism information from over 40 different publications into a coherent understanding of how ionizing radiations produce the killing of stem cells in human tumors will read somewhat like a monograph. Several physical and chemical parameters that can modulate the radiation response of clonogenic cells in tumors will be described according to their impact on the two mechanisms of cell killing.

The use of the LQ model in basic radiation mechanism studies with cells of relatively homogeneous radiation response is presented and will be extended to the fitting of survival data generated with heterogeneous cell populations (tumors) where the inactivation parameters are some combination of those for the various cells that make up the whole. The use of the LQ model for predicting normal tissue complication responses (where mechanisms other than stem cell killing could be active) will be briefly addressed. These mechanisms will be discussed more fully in the planned Volume II. No comprehensive review of all the molecular mechanisms that could impact the radiation treatment of human cancers will be attempted since, in most cases, these studies do not distinguish between the two mechanisms of cell killing by radiation. The potential molecular targets related to alpha- and beta-inactivation will be described along with suggestions for further molecular characterizations of these two independent processes. This textbook should read as an advanced "target theory."

While radiobiology research over the past three or four decades has produced meaningful parameters for medical physicists to initiate TCP and NTCP modeling that predicts for altered cancer treatment prescriptions, the efforts to define clinical assays that inform about clonogen radiosensitivity, tumor oxygenation and tumor growth fraction, in general, were not successful. Although predictive assay research was informative, no standard assays are currently utilized to obtain this information from individual patients. Consequently, several important parameters that must be input into the models are best estimates or "guestimates" obtained from various published studies. In spite of these shortcomings, the time has come to investigate in clinical trials all the potential benefits that might arise from the recent advances in cancer imaging and dose delivery. It is our hope that this book, focused on quantitative radiobiology in the LQ formulation, will assist both medical physicists and radiation oncologists to identify improved cancer treatments. And if it inspires current investigators to translate potentially improved radiotherapy schedules based upon TCP and NTCP modeling to actual patient benefit, our efforts in assembling this volume will not have been in vain.

The cover of this book is a reworking of a figure that was presented in a poster at the International Congress of Radiation Research in Dublin, Ireland, in 1999. It obviously says that the understanding of cell killing by radiation and its use in radiotherapy is like a jigsaw puzzle where different specialties are merged together to relate the underlying physics, chemistry, biophysics, biochemistry, pathology and clinical outcome data that are available. Reece Walsh, an 11th grader who is skilled in computer graphics, created its final version. Through this experience, he learned some of the university-level sciences involved in cancer radiotherapy and could proceed to a career in medical physics, since he is truly gifted in mathematics and the physical sciences.

Acknowledgments

Don Chapman was mentored over his career by four "giants" of the radiation research field: Douglas Cormack (MSc supervisor) in radiation physics, Ernest C. Pollard (PhD supervisor) in cellular biophysics, Gerald E. Adams (postdoctoral collaborator) in radiation chemistry and Cornelius Tobias (director of biology research at the Lawrence Berkeley Laboratory) in charged-particle biophysics. His understanding of radiation mechanisms in human tumor cells is largely a synthesis of all their teaching, which inspired his laboratory studies. Their friendship and instruction is gratefully appreciated.

He was surrounded in his laboratories by many brilliant scientists who offered their collaboration on different projects. These include Eleanor Blakely, Jack Boag, Aloke Chatterjee, Stan Curtis, Indra Das, Dave Dugle, Ed Englehardt, Alan Franko, Colin Gillespie, Clive Greenstock, Cam Koch, John Lyman, John Magee, Yoshi Matsumoto, Malcolm McPhee, Ben Movsas, Alan Nahum, Malcolm Patterson, Jim Raleigh, Rene Santus, Richard Schneider, Oliver Scott and Raul Urtasun. His students and research fellows who contributed to these studies include Ozar Algan, Siham Biade, David Herold, Renuka Iyer, John Matthews, Ron Moore, Matthew Parliament, Wilson Roa, Chuck Thorndyke and Joan Turner. He always had very gifted technical assistance (his "hands" in the lab), particularly from Matthew Fenning, Jane Lee, Bert Meeker, Tony Reuvers and Corinne Stobbe. Most of the research collated in this book was performed with radiation oncology funds provided by generous department chairmen; Dr. James Pearson (Edmonton) and Dr. Gerald Hanks (Philadelphia).

A sincere thank-you goes to Reece Walsh, with assistance from his father Mike, for reformatting several graphs presented in this textbook and for creating the book cover.

His best collaborator over the past three decades has been his wife, Beverly, who has improved his syntax and corrected many spelling mistakes.

Alan Nahum has worked with many highly able scientists. Among his many outstanding collaborators on radiation dosimetry he should like especially to mention John Greening (PhD supervisor), Hans Svensson, Pedro Andreo, David Rogers, Alex Bielajew, David Thwaites, Stan Klevenhagen, Charlie Ma (his first PhD student), Richard Knight, Paul Mobit, Frank Verhaegen and Sudhir Kumar.

Gordon Steel introduced him to the fascinating world of radiobiology. Subsequently, he worked with many excellent researchers including Steve Webb, Beatriz Sanchez-Nieto (his first postdoc), Diana Tait, David Dearnaley, Stefano Gianolini, Francesca Buffa, Colin Baker, Eva Rutkowska-Onjukka, Catharine West, Giovanna Gagliardi, Mauro Iori, Ruggero Ruggieri, Julien Uzan, Isabel Syndikus and Aswin Hoffmann. The several months he spent at Fox-Chase in 2002 working closely with Don Chapman were hugely significant in deepening his knowledge of experimental radiobiology.

Additionally, he wishes to acknowledge Philip Mayles, the former head of medical physics at CCC and coeditor of *Handbook of Radiotherapy Physics,* for his friendship and for turning a blind eye to time spent on book writing.

Both authors wish to express their appreciation to Luna Han of Taylor & Francis who showed great optimism, patience and support during this book writing venture. And special thanks go to Judith Simon for guiding this project to its conclusion.

About the Authors

J. Donald Chapman is currently performing in retirement consulting services to various radiation medicine commercial and academic entities. He obtained the degrees of bachelor of science in engineering physics and master of science in radiation physics from the University of Saskatchewan and a PhD in biophysics from the Pennsylvania State University. He performed postdoctoral research at the Gray Laboratory in England prior to directing research laboratories at Atomic Energy of Canada (Pinawa, Manitoba, 1968–1977), the Cross Cancer Institute of  the University of Alberta (Edmonton, Alberta, 1977–1990) and the Fox Chase Cancer Center, which was then affiliated with the University of Pennsylvania (Philadelphia, 1990–2002). His tenure at Atomic Energy of Canada included a 1-year sabbatical at the Lawrence Berkeley Laboratory in Berkeley, California.

His research contributed to the fields of hypoxic radiosensitizing drugs, nuclear medicine markers of viable hypoxic cells, mechanisms of photodynamic therapy and the killing of tumor cells by ionizing radiations. He has authored and coauthored over 200 scientific articles in several scientific journals and conference proceedings and taught radiobiology principles to medical physics graduate students and radiation oncology residents at the University of Alberta and the University of Pennsylvania. From 1984 to 1989 he served as associate director of research at the Cross Cancer Institute in Edmonton. For over 25 years he served on the Radiobiology and Tumor Biology Committees (as chairman for 15 years) of the Radiation Therapy Oncology Group (RTOG) sponsored by NIH to implement innovative radiotherapy protocols for cancer patient treatments. He has served on the editorial boards of numerous radiation research journals, has received several international research awards and, in 1991, was secretary-general of the 9th International Congress of Radiation Research in Toronto, Canada.

Alan E. Nahum is head of physics research at Clatterbridge Cancer Centre (CCC) near Liverpool, UK, and visiting professor at the Department of Physics, Liverpool University. He read physics at the University of Oxford and then did a PhD, on theoretical radiation dosimetry, at University of Edinburgh. After 4 years as a schoolteacher, he joined Institutionen för Radiofysik, University of Umeå in Sweden (1979–1985), as a postdoctoral researcher and lecturer. In the spring of 1983 he was a visiting research officer at the Division of Physics of the National Research Council of Canada, Ottawa. Between 1985 and 2002 he worked at the Joint Department of Physics of the  Institute of Cancer Research (ICR) and Royal Marsden Hospital (RMH), Sutton, UK. He worked for short spells at Fox-Chase Cancer Center, Philadelphia (where he met Don Chapman); at the Radiation Oncology Department, l'Arcispedale S. Maria Nuova, Reggio Emilia; and at the Radiotherapy Department, Rigshospitalet (Copenhagen University Hospital), before joining Clatterbridge Cancer Centre (CCC) in 2004.

For his doctorate he wrote a Monte-Carlo simulation code to explain the anomalous response of the Fricke ferrous sulfate dosimeter in megavoltage photon and electron beams. This led to the "track-end" correction of the Spencer–Attix cavity integral and subsequently to many other contributions on dosimeter response in kilovoltage and megavoltage radiotherapy beams. He has co-authored several UK and international codes of practice for reference dose determination in external-beam radiotherapy. From the late 1980s his interests expanded to include radiobiology; he developed a biomathematical model for tumor control probability incorporating interpatient variation in radiosensitivity and tumor volume, and heterogeneity in tumor dose. His research groups at ICR/RMH and at CCC have created the widely used BIOPLAN and BioSuite software packages for radiobiological analysis of radiotherapy treatment plans. He has edited/co-edited three books, including *Handbook of Radiotherapy Physics* (Taylor & Francis 2007), and authored/co-authored approximately 170 peer-reviewed papers, book chapters, and conference proceedings. He has set up several research workshops and numerous courses, including Radiobiology & Radiobiological Modeling in Radiotherapy, launched in 2006. His current focus is on radiobiologically guided treatment optimization through individualization of tumor prescription and fractionation.

List of Abbreviations

α: the coefficient of cell inactivation by the linear mechanism of the LQ equation

AECL: Atomic Energy of Canada Ltd.

AMU: atomic mass unit

β: the coefficient of cell inactivation by the quadratic mechanism of the LQ equation

BED: biologically equivalent dose

CCC: Clatterbridge Cancer Centre

CCI: Cross Cancer Institute

CT: computerized tomographic imaging

DMIPS: dynamic microscope image processing scanner

DSBs: double-strand breaks in cellular DNA

DVH: dose volume histogram

FACS: fluorescent activated cell sorting

FCCC: Fox Chase Cancer Center

FSU: functional subunits of tissue

GTV: gross tumor volume

G-value: the number of chemical entities formed by 100 eV of absorbed dose

LBL: Lawrence Berkeley Laboratory

LDH: low-dose hypersensitivity

LET: linear energy transfer, usually in units of kiloelectron volts per micrometer (KeV/μm)

MHMT: the multihit, multitarget equation for cell killing

MHST: the multihit, single-target equation for cell killing

MRI: magnetic resonance imaging

NTCP: normal tissue complication probability

OER: oxygen enhancement ratio

PE: plating efficiency

PET: positron emission tomography

PTV: planned treatment volume

RBE: relative biological effectiveness

RE: relative effectiveness

SBRT: stereotactic body radiotherapy

SF: surviving fraction

SHMT: single-hit, multitarget equation for cell killing

SHST: single-hit, single-target equation for cell killing

SPECT: single photon emission tomography

SRS: stereotactic radiosurgery

SSBs: single-strand breaks in cellular DNA

TCP: tumor control probability

TGF: tumor growth fraction

TR: therapeutic ratio of tumor response/normal tissue complications

WIF: Withers *iso*-effect formula

① Introduction

The treatment of human cancer by ionizing radiation has developed over the past century into one of today's most effective modalities for managing this disease. Along with surgery and chemotherapy, this type of treatment comprises the major clinical tools for treating the cancers that will be detected this year and well into the future. Currently, radiation treatment alone or combined with other modalities is the preferred therapy for over 50% of all cancer pathologies. The protocols for radiation treatments in current use for various cancers were established mainly by empirical clinical experience over several decades. These have attempted to achieve a high level of killing of clonogenic tumor cells within planned treatment volumes (PTVs) while producing minimal or acceptable damage to the normal tissues that will inevitably be exposed to some dose of radiation during the treatment. The efficiency of tumor response (cell kill) relative to normal tissue complications is called the "therapeutic ratio" (TR), although TR is usually not defined quantitatively.

Most current protocols prescribe 1.8–3 Gy of radiation each day for 5 days/week for a fixed, predetermined total dose (i.e., a fixed number of fractions) considered acceptable to the involved normal tissues on a population basis. Dose fractions are not usually given on weekends for reasons that are not biological (although there could be benefits for normal tissue repair) but rather are based on the logistics of staffing and funding radiotherapy clinics. For a given tumor type the prescribed doses and dose fraction sizes vary only slightly between different treatment centers and are usually very conservative; that is, they are lower than what might be tolerated, even though the *modus operandi* of radiation therapy is that "the higher the dose of radiation, the greater the proportion of tumor cells that will be eradicated." Low or high dose-rate radiation delivered by brachytherapy procedures is the preferred treatment for some tumors, and hyperfractionation (<1.8 Gy/fraction) and hypofractionation (>3 Gy/fraction) techniques have also been investigated, the latter undergoing something of a renaissance today.

Medical physicists have always been involved in planning the delivery of radiation dose to individual human tumors, but with the availability of modern radiation sources, improved imaging technologies (CT, MRI, PET, etc.) and faster computers, their role has grown in importance. In most treatment centers, the GTV and PTV are still delineated by the radiation oncologist, but it then becomes the task of the treatment-planning physicists (and dosimetrists) to devise the optimal strategy for delivering the prescribed dose (uniformly) to that volume. From a long experience with tumors at every anatomical site, this plan might be produced after only one attempt, but in many situations two or three iterations could be required. Mention should be made of the "inverse planning" of intensity-modulated radiotherapy (IMRT); here the computer algorithm does the iterations until the normal-tissue dose-volume constraints and tumor dose set by the planner are met. And since dose volume histograms (DVHs) have become an integral part of all treatment planning software, the radiation exposure of critical normal tissues can be known with a greater degree of accuracy. Thus, the physics of higher energy photon delivery utilizing precise digital imaging techniques and high-speed computers has been integrated so that defining a uniform radiation dose across most PTVs is no longer a limiting step in the patient's treatment. The physics of the delivery of a prescribed radiation dose to PTVs has evolved to a very high level of certainty. Now is an appropriate time to inquire about the radio-responsiveness of the different cellular targets within PTVs that require inactivation for producing local control and tumor cures.

Starting around the 1950s, radiobiology research was performed to understand basic mechanisms associated with mammalian cell killing and mutation and to investigate the radiation sensitivities of the cells in various normal tissues. Most of that early research was funded by atomic energy agencies that were analyzing the devastating effects of the bombs dropped on Hiroshima and Nagasaki at the end of the Second World War. And since the construction of nuclear power plants for electrical energy production was in full swing, their potential risks to surrounding communities was of great importance. A component of this research was performed with cells irradiated as *in vitro* tissue cultures where most of the physical and chemical parameters that can modify intrinsic radiosensitivity could be rigorously controlled. These studies were augmented by research performed with multicellular spheroids, with animal tumors, with human tumor *xenographs* growing in rodent hosts and with animal normal tissues *in vivo*. This large compendium of experimental data (reported in thousands of peer-reviewed papers) constitutes the scientific underpinning of the fractionated clinical protocols employed today. These are summarized as the 4 or 5 Rs of radiotherapy (Withers 1975; Deacon, Peckham and Steel 1984; Steel 2007b). ***Repair*** of sublethal damage in normal tissues, along with the ***redistribution*** and ***reoxygenation*** of clonogenic tumor cells, is believed to be advantageous for maximizing tumor response. The continued growth of tumor tissue during a course of treatment (***repopulation***) is potentially detrimental for achieving the desired response and should be accounted for.

The intrinsic *radiosensitivities* of the clonogens in different tumor pathologies were found to be significantly different and correlated well with the responsiveness of tumors observed in clinical studies (Deacon et al. 1984; West et al. 1993, 1997). From the 1980s onward, most of the funding for radiobiology research was derived from those agencies that had been created to address cancer research, cancer patient treatment and risks of radiation exposure during space travel.

It has been well established that the killing of tumor cells by radiation involves molecular damages produced in cellular DNA (Steel 2002), the molecule that contains the genetic template essential for cell viability and reproduction. The different types of damage are molecularly diverse and can be produced by direct ionizations arising from charged particle interactions with the target molecule and/or by various free radicals produced in cellular water that diffuse to and react with the target molecule. And since over 70% of a human cell (by weight) consists of water, the energy deposited in that cell constituent can be expected to have major effects on the integrity of cellular DNA. In fact, the indirect effects of diffusible OH$^•$ radicals produced by the radiolysis of cellular water account for at least 65% of the cell killing observed in aerobic mammalian cells (Chapman et al. 1973, 1979). The identification of water radiolysis products and their subsequent reactions with important cell molecules provided a wealth of information essential for our current understanding of cell killing by radiation and for its manipulation for therapeutic gain (Adams 1972; Ward 1975). The elucidation of important radiation chemistry pathways that linked the physical events of radiation deposition in cells to potentially lethal molecular damage in DNA constituted a most important phase of radiobiology research (Chapman and Gillespie 1981). It also indicated potential strategies for improving conventional cancer treatments.

Unfortunately, most of the alternative strategies for improving clinical response based upon laboratory research resulted in no or only marginal benefit to patients, and some strategies actually proved harmful. The early clinical trials with neutron radiation were based upon inappropriate values of relative biological effectiveness (RBE) that had been determined with higher dose fractions than those used in patients and produced unacceptable normal tissue toxicities (Stone 1984). We now know that the RBE of high linear energy transfer (LET) radiation increases inversely with dose fraction size since increased cell killing is mainly due to increased single-hit inactivation (Chapman, Blakely, et al. 1978). Thus, the normal tissue complications associated with neutron therapy might have been predicted. The combination of hyperthermia with radiation and the use of hypoxic cell radiosensitizers did not produce the expected clinical benefits. Nevertheless, the large base of associated laboratory studies that accompanied these approaches can now explain why they failed to improve treatment outcomes. The hyperthermia enthusiasts found it extremely difficult to deliver the high uniform temperatures across human tumor PTVs—those that had demonstrated efficacy in animal studies. The hypoxic radiosensitizer enthusiasts were unable to deliver to patients at each dose fraction the high drug doses that showed effectiveness in animal tumor models, due to toxicity limitations.

Thirty years ago several schools of radiation oncology taught that the difference between the intrinsic radiosensitivity of the clonogens in different human tumors was not large enough to account for the significant difference observed in their clinical response. It has now been demonstrated that the radiation killing of human tumor cells *in vitro* after doses of 2 Gy (SF_{2Gy}) is variable and correlates well with *in vivo* tumor response to fractionated and brachytherapy treatments (Steel and Peacock 1989; West et al. 1993, 1997).

Radiation and medical biophysicists have always attempted to describe the killing of tumor cells by radiation in terms of mathematical expressions that relate radiation dose to lethal molecular damages. For years, the single-hit, multitarget (SHMT) equation (a two-parameter model) was the "workhorse" of this field of quantitative radiobiology, and the terminal slopes and their extrapolation numbers with zero dose were considered to contain important mechanistic information. Radiobiologists who measured the effects of high LET radiations on cell killing were first to combine a simple exponential function with the SHMT equation (yielding a three-parameter model) to describe the effectiveness of various heavy charged particles. But over the past 30 years, quantitative radiobiology has cautiously adopted the linear-quadratic (LQ) equation (a two-parameter model) to describe tumor cell killing. Unfortunately the LQ model in widespread use today is frequently misunderstood, misapplied, or used outside its region of validity (Chapman and Gillespie 2012; Chapman 2014). The LQ equation can accurately describe the radiation killing of human tumor cells if they are of homogeneous radiosensitivity and when the radiation is administered under conditions where there is no repair of sublethal damage during the exposure. For all other irradiation conditions (which have been employed for the majority of the published investigations) the basic LQ expression should be modified to account for the violations of these strict criteria. In fact, there are at least four different LQ equations required to describe radiation-induced cell killing *in vitro* and *in vivo*. Where and how these should be utilized is described in this textbook. For example, survival data produced with asynchronous cells will yield values of α and β that are complex associations of the parameters unique to cells in the various stages of the cell-division cycle that compose the population (see Chapter 3). We suggest that the inactivation parameters derived from heterogeneous cell populations be distinguished from those of homogeneous populations by placing a bar over the symbols. Further, when radiation dose-rates of 1–2 Gy/min are utilized with cell exposures at room temperature or at 37°C, there will always be less cell killing than predicted by the basic LQ model (Gillespie et al. 1976; Chapman and Gillespie 1981), especially at the higher doses (see Chapter 5). Fortunately, experimental data for cell killing under these conditions have been published and a modified LQ expression that includes a rate of sublethal damage repair can quantitatively account for the reduced cell killing that is observed (Chapman and Gillespie 1981). When radiotherapy is administered in daily fractions, the LQ survival curve will be reinitiated each day (we usually assume identical inactivation parameters and complete repair of sublethal damages) to produce a near exponential survival over several decades of tumor cell killing. This response requires yet another expression, which follows from the LQ equation (see Chapter 10). It is hoped that the description of the LQ model in this volume will encourage investigators to declare how they intend to use it and to take into account the potential complicating factors.

Chapman has utilized the LQ model for over 30 years to characterize the effects of both physical and chemical factors known to modulate the intrinsic radiosensitivity of mammalian (including tumor) cells. Most of that research was performed with relatively homogeneous populations of G_1-phase and G_0-phase (both are diploid) or mitotic (tetraploid) cells irradiated at 4°C–6°C, where sublethal damage repair is minimized. Chapters 2–5 describe the application of the LQ model for the biophysical understanding of the two independent mechanisms by which tumor cells can be inactivated. In Chapter 3, the modifications to the LQ model to produce meaningful values for single-hit and double-hit inactivation parameters for cell populations of known mixed intrinsic radiosensitivities are described. This modified equation should be used for studies that utilize asynchronous populations of tumor cells, for mixtures of aerobic and hypoxic cells and for mixtures of quiescent and proliferating cells. The parameters $\bar{\alpha}$ and $\bar{\beta}_0$ will be used when the LQ model is fitted to survival data generated with cell populations of heterogeneous radiosensitivity.

Chapter 6 will describe the microdosimetry and nanodosimetry of energy deposition in cells of tumors and normal tissues and how these physical events interact with DNA targets to produce the cell death whose kinetics are linear and quadratic with respect to radiation dose. When linear and quadratic parameters are extracted from data sets of human tumor response and normal tissue complication induction, the results can have empirical utility but are unlikely to inform about underlying cellular mechanisms. But this limitation need not dampen our enthusiasm for testing different fractionation schemes since our current clinical protocols were also derived empirically and are extremely effective. When cell killing is the basis for the radiation effect being investigated, the LQ model should provide insight as long as the various factors that can modify radiosensitivity are accounted for. Unfortunately for both tumor control probability (TCP) and normal tissue complication probability (NTCP) modeling, the biologically relevant details of the tissues under treatment are usually not known with sufficient accuracy. After three decades of gallant attempts we still do not have practical and reliable predictive assays for tumor cell intrinsic radiosensitivity, for tumor hypoxic fraction, or for the proportion of proliferating versus quiescent cells (Chapman, Peters and Withers 1989). Modeling tumor and normal tissue response to various radiation treatments can always be improved as the specific biology of the tissues in question becomes better known.

The laboratories of Chapman produced over 3,000 survival curves (each consisting of 7 to 10 survival points) for mammalian cells irradiated under numerous physical conditions and chemical environments. These data are published in over 40 different peer-reviewed manuscripts in several different journals and are amalgamated here to form the backbone of the quantitative radiobiology presented in this textbook (Chapman 1980; Chapman and Gillespie 1981; Chapman 2003). Chapman's research "habits" were formed under the mentoring of Doug Cormack, Ernest C. Pollard, Gerald Adams and Cornelius Tobias, whose many contributions to our understanding of radiation mechanisms are well known. Many of the studies were designed as research projects for radiation oncology residents as part of their academic training and funded with departmental funds provided by generous clinical directors (Dr. Jim Pearson at the Cross Cancer Institute in Edmonton and Dr. Gerald Hanks at the Fox Chase Cancer Center in Philadelphia). The studies performed at the Lawrence Berkeley Laboratory were facilitated by an NIH program project grant of which Professor Tobias was the principal investigator.

Ernest C. Pollard, Chapman's PhD mentor, cautioned students and junior colleagues about becoming too "enamored" with the mathematical modeling of radiation mechanisms in cells. Pollard, who had studied for his PhD under James Chadwick at the Cavendish Laboratory in Cambridge, England (at the time of the discovery of neutrons), had missed out on an important discovery by "wasting months in believing theoretical estimates of a novel isotope decay rate instead of measuring it directly in the laboratory." He taught his students that "with two parameters you can explain most science; with three parameters you might learn more but only if you can control each independently for experimental investigation, and with four parameters you can make an elephant walk." This advice probably applies to those who wish to quantify the effects of ionizing radiation on human tumor and normal cells, especially when fitting mathematical models to cancer treatment outcomes. The modeling of cell killing by more complex expressions involving additional parameters makes little sense unless each parameter can be independently validated through laboratory research. Additional parameters in cell killing expressions can result in better data fits but usually with a loss of precision and understanding of each parameter.

Alan Nahum's first contribution to radiobiology was the development of a mechanistic model of TCP in the early 1990s (Nahum and Tait 1992; Webb and Nahum 1993) while at the Institute of Cancer Research (ICR)/Royal Marsden Hospital, UK. His first radiobiology mentor was Gordon Steel. The "Marsden" TCP model was the first to employ radiation parameters with a distribution in the α-inactivation parameter to account explicitly for interpatient heterogeneity (Nahum and Sanchez-Nieto 2001). This produced much "flatter" TCP versus total dose curves that were similar to those obtained in clinical practice (see Chapter 10). It was now possible to obtain clinically realistic, sigmoidal local control versus dose curves using appropriate input parameters for intrinsic tumor cell radiosensitivity along with reasonable estimates for clonogen number. Their TCP model also incorporated the effect of non-uniform doses through the differential DVH that is becoming available in three-dimensional (3-D) treatment planning systems. At the same time, the first papers on NTCP modeling began to appear—for example, the Lyman–Kutcher–Burman NTCP model (Lyman 1985; Kutcher and Berman 1989) together with endpoint-specific parameters (Emami et al. 1991). Nahum's research group understood immediately the potential of *iso*-NTCP (or "isotoxic") customization of the tumor dose (Nahum and Tait 1992). Ever since those heady 1990s Nahum has maintained a strong interest in treatment plan optimization based on radiobiological models. His postdoctoral research group member, Beatriz Sanchez-Nieto, wrote the BIOPLAN software for computing TCP and NTCP (Sanchez-Nieto and Nahum 2000) and further developed the idea of isotoxic prescription dose individualization (Sanchez-Nieto et al. 2001). John Fenwick (Nahum's PhD student) developed a model for rectal complications (Fenwick thesis) and Francesca Buffa (another PhD student) reanalyzed Catharine West's unique cervix SF_{2Gy} data set using the Marsden TCP model (Buffa et al. 2001). After moving to the Clatterbridge Cancer Centre (CCC) in 2004, Nahum has gone on to propose that doses to non-small-cell lung tumors be "customized" according to NTCP-based estimates of the risk for radiation pneumonitis. He and Julien Uzan have developed the BioSuite software to customize "isotoxicity" for both dose and number of fractions (Uzan and Nahum 2012); BioSuite makes predictions consistent with the current SBRT protocols for extreme hypofractionation of stage T1/2 NSC lung tumors. The CCC group has extended its interests to radiobiological inverse-planning- and functional-image-based "dose painting" (Uzan and Nahum 2012).

This textbook is the result of collaboration between Don Chapman and Alan Nahum that started over 10 years ago when they were at the Fox Chase Cancer Center in Philadelphia, Pennsylvania. Their first efforts resulted in an analysis of the probability of controlling prostate tumors incorporating the then-new measurements of frequent and severe hypoxia in a proportion of tumors in men for whom brachytherapy had been prescribed (Nahum et al. 2003). This analysis questioned the proposition that prostate cancer clonogens have a low alpha/beta ratio (Brenner and Hall 1999). It also pointed out how widely different input parameters to the TCP model can produce equally good fits to clinical data and underscored the importance of knowing with some certainty as much tumor biology information as possible about the specific tumors of interest. The Chapman–Nahum collaboration was a true "marriage of two minds" bringing different but complementary skills to the table: Chapman's hands-on knowledge of quantitative cellular radiobiology and Nahum's clinically relevant biological modeling skills. Marco Schwarz has investigated how NTCP-based plan individualization can improve cancer treatments. He will collaborate on Volume II of this textbook that will describe TCP and NTCP modeling techniques in more detail and the use of these models in treatment-plan optimization (Uzan and Nahum 2012).

The collaboration of the authors has been reinforced by the introduction of a teaching course, Radiobiology and Radiobiological Modeling in Radiotherapy, organized by the CCC, Wirral, UK (where Nahum is employed) that covers most of the material presented in this book. The course has been offered each year since 2006 by a 20-strong faculty, each an expert in his or her own field. It has been supported by those companies that are enthusiastic about incorporating radiobiological expressions, tumor biology parameters, and maybe models for predicting tumor control and normal-tissue complications into frontline treatment planning for specific human cancers.

It is the authors' conviction that this area of radiation oncology is ripe for additional clinical research. Novel protocols of radiation treatment (involving altered fraction size and treatment times) that will maximally exploit the improved conformality of dose distributions obtained with protons, carbon ions, gamma-knives, tomotherapy and other modern delivery technology within better defined PTVs (obtained with new imaging procedures) await testing in clinical trials. Again we note that the majority of our current radiotherapy strategies were derived by empirical optimization of clinical experience performed with inferior technologies. The "retuning" of these procedures based upon state-of-the-art tumor imaging and dose delivery systems could produce equally good or better treatment outcomes with less normal tissue trauma (Thiagarajan et al. 2012). And if larger dose fractions can be safely administered daily or even once, twice, or three times a week, the impact on patient stress (less frequent clinic visits and patient setups) and health care costs (the patient setup time is typically much longer than the beam-on time) could be enormous.

This textbook on quantitative radiobiology is intended as a resource for those who attempt to compute *iso*-tumor and *iso*-normal tissue responses in novel clinical protocols. It can be seen as a practical manual on how to take the information gleaned from quantitative radiobiology over the past 40 years and apply it to design altered and improved cancer treatments.

The emphasis on quantitative radiobiology in this text is based upon our belief that the parameters for tumor cell intrinsic radiosensitivity and other tumor biology parameters that are input into tumor response models are of utmost importance. The minimal mathematics required to define precisely (and investigate mechanisms of) cell killing is described. And since traditional tissue culture experimentation is somewhat passé (not fundable), we may have to make the most of the best quantitative information currently in the literature. This book should also instruct those that participate in molecular biology research about the two independent mechanisms of cell killing that should be addressed in their studies. The factors and molecular mechanisms associated with single-hit killing are likely to be very different from those associated with double-hit killing. Procedures for investigating mainly single-hit or double-hit killing are described in the book. We believe it will be useful for teaching radiobiology to medical physics treatment planners and radiation oncology residents for at least the next decade.

② The Generation of Quantitative Radiobiology Data

This chapter will introduce the basic linear-quadratic (LQ) equation as the most appropriate and least complicated model for describing cell inactivation by ionizing radiation. The physical conditions required during radiation exposures for cell killing to conform to the LQ model are identified. That these conditions are poorly understood and rarely met in many published studies suggests that arguments about LQ appropriateness will arise. The basic LQ model obtains for cell populations of homogenous radiosensitivity when irradiated under experimental conditions that eliminate or minimize the repair of sublesions during the radiation exposures. Consequently, survival curves produced with asynchronous populations of cells irradiated at dose-rates of about 1 Gy/min at room temperature or at 37°C will yield α- and β-values from which information about fundamental biophysical mechanism(s) will be difficult to interpret.

The inactivation of cells by ionizing radiation was initially investigated by physicists (in the early 1900s) who were concerned about the potential detrimental effects of these novel rays on living systems and who were also interested in their exploitation for medical benefit. Most of the early researchers had training in atomic or health physics and many were mathematically inclined. Consequently, there was great interest in understanding mechanisms of radiation effects on cells through mathematical models. The age of "target theory" burgeoned during the middle portion of the last century and radiation killing was postulated to result from **lethal damage(s)** produced in **molecular targets** by **specific ionizing events**. Much of that early research was devoted to defining the most important radiation deposition events and the molecular targets of consequence. Several textbooks that relate to these efforts have been published, including those by Lea (1962), Elkind and Whitmore (1967), Okada (1970), Alper (1979), Chadwick and Leenhouts (1981), Thames and Hendry (1987), Steel (2002) and Hall and Giacca (2006). One of the best descriptions of the mathematical models for cell killing by radiation is in *The Radiobiology of Cultured Mammalian Cells* (Elkind and Whitmore 1967). The development of single-hit/single-target (SHST), single-hit/multitarget (SHMT), multihit/single-target (MHST) and multihit/multitarget (MHMT) formulations is presented. These equations utilize Poisson statistics to compute the fraction of a cell population that will escape killing (will proliferate to form a colony) after specific radiation exposures whose ionizations are defined as random events. For many viruses, bacteria and yeast, radiation killing follows a simple exponential function defined by the SHST equation. For mammalian cells whose radiation inactivation displayed a downward curvature with increasing dose, the SHMT became the preferred expression. Radiation research meetings were filled with presentations that described specific ionizing events and hypothesized about what constituted the molecular target or targets. This quantitative approach to understanding the effects of radiation on living cells has served the field well and assisted with many interlaboratory comparisons of effects, although it can be said, in hindsight, that it was a relatively inefficient system. As more and more factors that influenced cell killing by radiation were identified, researchers came to wish that they had performed last week's experiments somewhat differently. And to this day, several investigators still choose to describe their mammalian cell inactivation data in terms of the SHMT equation:

$$SF = 1 - (1 - e^{-k_1 D})^n \tag{2.1}$$

where

 SF is the fraction of a cell population that survives a radiation dose of D
 k_1 is the terminal exponential slope on a semilogarithmic plot of log SF versus D
 n = the number of targets per cell that require inactivation and is equal to the intercept of the terminal slope with the zero dose axis

Figure 2.1 shows survival data for stationary phase (G_0) Chinese hamster fibroblast cells (V-79-379A) irradiated with 250 kVp x-rays at low temperature (to inhibit the repair of sublethal damages during the radiation exposure). The surviving fractions were derived from enumerating over 1,000 surviving cells (measured as colonies of >50 cells after 7 days of postirradiation incubation) for each radiation dose—a number large enough to ensure that the statistical (binomial) error of each data point was less than the size of the graphical symbol. Of course, cell handling procedures (culturing, trypsinization, diluting, pipetting and colony counting) would also contribute to the intrinsic experimental error of each survival point. A computer fit of these data to the SHMT equation (the dashed line) shows that cell killing is underestimated in both the low-dose and the high-dose regions. Over the survival range that is easiest to measure, from SF of 0.5 to 0.01 (radiation doses from 3 to 8 Gy), the model gives a reasonable estimate of cell killing. But, since the equation does not yield an accurate estimate for cell killing at 1.8–2 Gy, the dose fraction size of greatest clinical relevance,

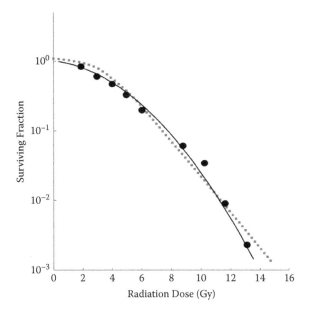

Figure 2.1. The surviving fraction of stationary-phase Chinese hamster fibroblasts (V79 cells) after various doses of 250 kVp x-rays. The cells were held at ~5°C to minimize the repair of sublethal lesions during the radiation exposure. Each survival point was determined with over 1,000 colonies so that the binomial error would be less than the size of the symbol. The dashed line is the best fit to the single-hit, multitarget (SHMT) equation and the solid line is the best fit to the linear-quadratic (LQ) equation.

several investigators considered that the SHMT equation would not be the best for modeling the response of human tumors after fractionated treatments.

During the 1970s, alternative mathematical descriptions of cell killing by radiation were generated and discussed. A physical (microdosimetry) approach employed specific energy density (δ) to account for cell killing and resulted in the **theory of dual radiation action** with its LQ formalism (Kellerer and Rossi 1972). From a molecular target (biophysical) approach, Chadwick and Leenhouts (1973) suggested that the lethal radiation lesions in mammalian cells were unrepaired DNA lesions that could be produced by both single-particle (proportional to D) and two-particle (proportional to D^2) radiation events. Discussions at meetings of microdosimetry were often heated, with the radiation physicists insisting that the physical events were the defining factor while the more biologically inclined persons insisted that the structure of the molecular targets within cells played the more important role. And, then, participants like Gerald Adams, Larry Powers, John Ward, Don Chapman and others, at the time, were investigating the radiation chemistry "interface" between the physical events and the molecular damages produced in cells. They were inclined to take a more "intermediate" position. In hindsight it is fair to say that neither the purely physical nor the DNA damage approaches of the 1970s were completely accurate but this research certainly prompted the field of quantitative radiobiology to investigate the merits of the LQ expression. And from the early 1970s onward, the Chapman laboratories became one of the strongest proponents of the LQ model, mainly because it was the only two-parameter equation that could describe their cell survival data with statistical precision (Gillespie et al. 1975). For example, when a "best fit" to the LQ equation was generated for the data in Figure 2.1, the resultant curve (the solid line) described very well the cell killing observed in the low-dose, intermediate-dose and high-dose regions. Again, it should be emphasized that the irradiations were performed with G_0-phase cells (of relatively homogeneous radiosensitivity) at low temperature, where little or no repair of the sub-lesions associated with the quadratic mechanism would occur during the radiation exposure.

The basic LQ equation of tumor cell killing is the product of two "independent" Poisson escape probabilities from lethal events produced in cellular targets according to first-order and second-order dose kinetics. It was always assumed that the physical events that produced cell killing could be qualitatively distinct and that their molecular targets could be different, resulting in a more complex target theory.

$$SF = (e^{-\alpha D}) \times (e^{-\beta_0 D^2}) \qquad (2.2)$$

where

SF = the fraction of cells that survive a radiation dose of D

α = the rate of cell killing by single-ionizing events

β_0 = the maximal rate of cell killing by two ionizing events observed when there is no repair of sublethal lesions during the radiation exposure

When survival data are best fit to Equation (2.2) by minimizing the sum of squares of deviations, the estimates of α and β_0 produce "confidence contours" from which estimates of error can be obtained (Gillespie et al. 1975). This equation then describes cell killing from two independent processes produced by single-hit and double-hit physical mechanisms, respectively. **The LQ model does not directly yield information about the microdosimetry associated with the energy deposition events nor does it inform about the nature of the resultant molecular lesions**. It was additional research (described mainly in Chapter 6) that produced an improved understanding of both the physical events and an indication that different molecular targets and lesions were associated with these mechanisms.

There have been suggestions that the LQ equation might be simply a truncated version of the true survival equation, which contains higher powers of radiation dose (D^3, D^4, etc.). Since it was extremely difficult to perform laboratory experiments that yielded good estimates ($\pm 10\%$) for just the two parameters, α and β_0, it was considered that the fitting of such data sets to an equation with additional parameters was unlikely to yield a greater understanding of mechanisms. Furthermore, these two parameters yielded statistically valid values (tested by χ_s^2 error analysis) for the majority of survival curves (Gillespie et al. 1975).

Other cell-killing models that have been investigated over the past few years include the repair–misrepair (RM) (Tobias et al. 1978), the repair saturation (Goodhead 1985) and the lethal, potentially lethal (LPL) damage (Curtis 1986) models. These are discussed briefly in Chapter 5, where sublethal damage is addressed and its repair is factored into the LQ model. In some cases, these models include additional parameters that were not amenable to laboratory measurement and consequently would be difficult to validate by experiments. It should be noted that this basic LQ equation always produced good fits to experimental data when rigorous experiment procedures were employed.

Firstly, the cell population under investigation should be of homogeneous radiosensitivity. To abide rigorously by this caveat would require that all *in vitro* radiobiology experiments be performed with cells that had been synchronized to a known cell cycle position where they exhibit a singular intrinsic radiosensitivity. And even if this could be achieved routinely, different cells in the exact cell cycle position might exhibit some variation of radiation responsiveness. Chapter 3 of this book describes the significant differences in α- and β_0-inactivation rates measured for hamster fibroblast and human tumor cells as they traverse the cell-growth cycle. These inactivation rates are likely to be averages of radiosensitivity distributions, albeit with much smaller variance than those of asynchronous cell populations, and will depend upon the goodness of cell synchronization. Several studies have utilized contact-inhibited cell populations (see Figure 2.1) since they reside in a stationary (non-proliferating) phase of the growth cycle (G_0) that is post-mitotic with a diploid complement of DNA, similarly to G_1-phase cells. In solid tumors, most clonogenic cells are "out of cycle" (quiescent) at the time of radiation treatment. The proportion of proliferating (in cycle) to total (proliferating and quiescent) cells is defined as the tumor growth fraction (TGF) and is usually less than 10% for most human tumors. The vast majority of cells in normal tissues are not proliferating, are differentiated and are in the G_0-phase. Thus, non-dividing cell populations are likely to express a more homogeneous response to radiation than those that are undergoing mitotic cell division. But the majority of published radiobiology research has utilized asynchronous populations of growing cells since they are the most readily available. Whether or not the resultant values of α and β_0 from these studies (complex averages of different sensitivities expressed by the various subpopulations) have mechanistic meaning will depend on the specific research question of interest (see Chapter 3). In spite of this fact, the differences in α-inactivation values measured for several tumor cell lines is significantly greater than the variations observed for any one cell line during the interphase of its cell cycle (Biade, Stobbe and Chapman 1997; Chapman 2003). For studies of radiation mechanisms in particular, the most informative values of α- and β_0-inactivation will be generated with cell populations of relatively homogeneous radiosensitivity.

The second caveat for producing accurate values of α and β_0 is that there be little or no repair of the sublesions associated with β-inactivation during the radiation exposure. This poses a problem for many radiobiologists since the radiation sources available to their research usually have dose-rates of 1–2 Gy/min at convenient radiation positions. Consequently, the exposure times required to investigate cell killing at higher doses will be several minutes and, at room temperature or at 37°C, sublesion repair during these exposures will result in higher values of SF. When such data are fitted to the simple LQ equation, they will yield β-inactivation rates that are lower than maximal (and the value of α will usually be higher as a consequence). This problem can be largely circumvented by irradiating cells at low temperature to inhibit the action of repair enzymes (say 4°C–6°C, a procedure adopted by the Chapman laboratories) or at dose-rates of >10 Gy/min. Our research at AECL and the CCI had access to both 250 kVp x-ray (1–2 Gy/min) and high dose-rate Co^{60} gamma (20–170 Gy/min) sources. The Co^{60} gamma irradiators were mainly for radiation chemistry studies, but cell killing by both sources at 4°C and at 37°C was measured to confirm the importance of both temperature and dose-rate on values of α- and β-inactivation parameters obtained from the survival curves generated. The majority of cell-survival data presented in this textbook was produced with cells irradiated in slowly stirred suspension cultures at low temperature. It was always recognized that the photon energies, dose-rates and temperatures of concern in clinical radiotherapy would be significantly different.

Figure 2.2 shows a glass irradiation chamber designed and fabricated at the LBL in 1975 that facilitated the irradiation of cells in slowly stirred suspensions equilibrated with different gases (oxygen, nitrogen, carbogen, etc.) and other chemical environments (hypoxic radiosensitizers and various protectors) in track segments of charged-particle beams produced at the BEVALAC. It could be positioned in a water bath at temperatures from 4°C to 50°C. Dose fractions of predetermined size were administered to the cell suspensions and samples for colony formation analysis were removed through the long-arm port, a procedure that guaranteed that the same cell population was being sampled after each radiation dose. Cells that had been synchronized by the mitotic shake-off procedure and incubated into various

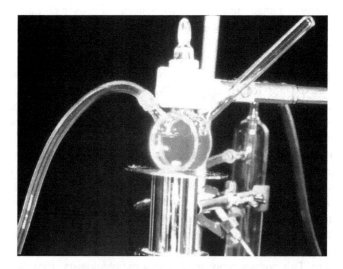

Figure 2.2. The glass irradiation chamber that was designed and first constructed at the Lawrence Berkeley Laboratory to facilitate the irradiation of asynchronous and synchronized cell populations under hypoxic, oxygenated and chemically modified conditions in various track segments of heavy-charged particle beams produced at the BEVALAC. The chamber could be immersed in a water bath at various temperatures and cells were sampled through the long-arm port after sequential radiation exposures.

positions of the cell cycle could be placed into the suspension chambers at reduced temperature to impede progression through the cell cycle for radiation characterization. Radiation doses delivered to the sample volume of this chamber were measured by Fricke dosimetry and were time averages of the exposures over the whole volume of the chambers. Using this device, seven- to nine-point radiation survival curves were generated with Chinese hamster fibroblast and human tumor cell lines under several conditions of physical and chemical modification to determine their effects on the single-hit and double-hit mechanisms. The majority of data presented in this book was generated using this irradiation chamber that was later copied (by excellent glass blowers) at both the CCI and the FCCC.

One interesting feature of the LQ equation is that it can be written as

$$-[lnSF]/D = \alpha + \beta_o D \tag{2.3}$$

When the data of Figure 2.1 are plotted as $-[lnSF]/D$ versus D (see Figure 2.3), radiation response is well described by a straight line with α as its intercept with the y-axis and β_o as its slope.

If the cells had been exposed to radiation at 37°C, the high-dose points would exhibit significantly reduced values, indicative of repair of sublesions during the radiation exposures. If a straight line is then fitted through such data, it will have a larger value of α and lower value of β_o than for the same cells irradiated under conditions of no or minimal repair. This method of plotting survival data can be useful for visually comparing different survival curves to determine if experimental factors operate on α- or β_o-parameters differently. We reiterate that repair must be minimized when generating cell survival data from which precise values of α- and β_o-inactivation are anticipated. This criterion becomes most

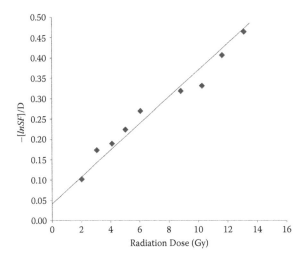

Figure 2.3. The surviving fractions of Figure 2.1 replotted as $-[lnSF]/D$ versus dose. The best-fit linear regression through the data points yields $\alpha = 0.045$ Gy^{-1} and $\beta_o = 0.033$ Gy^{-2}.

important for investigation of biophysical mechanisms. The matter of dose-rate and exposure times will be further discussed in Chapter 5.

When interlaboratory comparisons of data are to be made, a third criterion for generating useful values of intrinsic cell radiosensitivity is that similar chemical and physical conditions be employed. Aerobic radiosensitivity (cells equilibrated with air at the time of radiation) is most frequently investigated since it is the easiest to reproduce. Rendering cells hypoxic (oxygen free) before radiation exposure is not a simple task (see Chapter 6). With regard to standard physical conditions, different research groups perform their studies with different radiation sources. The literature includes studies performed with x-ray sources of 50 to 300 kVp maximal energy, with Cs^{137} and Co^{60} γ-ray sources and some with linear accelerator radiations of even higher energies. These different sources will produce different secondary electron spectra at the position of cell irradiation, which could impact the α- and β_o-inactivation parameters. Some studies indicate that x-ray sources of the lowest energies have slightly larger α-inactivation parameters than those produced by higher energy photon sources (Reniers et al. 2008; Hill 2008). In general, this effect is relatively small (less than 20% difference) and has not contributed to serious interlaboratory problems.

In many studies, mammalian cells are irradiated as monolayers while attached to plastic or glass surfaces. The preference of the Chapman laboratories to irradiate cells in suspension cultures allowed for maximal flexibility for studying synchronized cells, for producing hypoxia and other chemical environments, for regulating temperature and for minimizing the effects of electronic buildup. When cells are irradiated attached to glass surfaces (or materials of higher Z-value) during radiation, backscattered electrons comprise a significant fraction of the dose delivered to the cells and radiation quality factors could come into play (Chapman, Sturrock, et al. 1970; Zellmer et al. 1998). These effects are described in Chapter 6 along with the potential clinical role for additional dose scattered from high-Z materials implanted into tumors.

These, then, are the most important factors that should be controlled when generating survival data with mammalian and tumor cells *in vitro* from which accurate values of α and β_o are derived. Most published studies describe the inactivation of asynchronous populations of cells irradiated at room temperature with some exposure times longer than 5 min. In addition, the radiation sources employed are of significantly lower photon energy than those used for most cancer radiotherapy. The relevance of such data to current clinical practice is not entirely straightforward.

It is important to understand how mammalian cell death after irradiation *in vitro* is measured and defined. While cell proliferation assays have become the workhorse for investigating cancer drug mechanisms (Weisenthal and Lippman 1985; Brock, Maor and Peters 1985), the "gold standard" for cell killing by radiation remains the **colony-forming assay.** This technique entails irradiating tumor cells either in suspension culture or attached to material surfaces, placing known quantities into pertri dishes (often requiring an enzyme procedure to release the cells from the material surfaces followed by dilution procedures) and incubating for several days at 37°C in atmospheres of air + 5% CO_2 until the individual colonies arising from single surviving cells can be observed by eye. Cell colonies are usually stained with some protein dye (e.g., trypan blue) before their number is enumerated. Mammalian cell colonies become visible to the eye when they consist of about 50 cells, requiring about five to six cell divisions. Cells undergoing no or fewer divisions (that are not visible) are considered to be dead. The number of colonies can be enumerated either manually or by electronic imaging devices and compared to the original number of cells that were plated. The plating efficiency (PE) of any cell population is defined as the number of colonies observed/number of cells plated. It should be noted that the PE of unirradiated human tumor cell lines is usually in the range of 0.5 to 0.9; that is, not all of the cells that you place in a culture dish will produce a visible colony. By dividing the PE_x for cells irradiated (dose = x) by the PE_0 of unirradiated (control) cells, values for surviving fraction (SF_x) can be generated for each radiation exposure. These values of SF are then input into computer programs to determine the best-fit values of specific radiation parameters from the various mathematical models. And as will be discussed in Chapter 7, the *in vitro* PE of unirradiated cells released (by enzyme procedures) from primary tumor biopsies is only 0.05%–0.5% or less; that is, the vast majority of the viable cells that are plated never form colonies. The colony-forming cells derived from such tumor biopsy specimens might constitute those that are defined as "tumor clonogens" or stem cells.

Mammalian cells can die by two quite different processes after exposure to ionizing radiations. Early research with Chinese hamster fibroblast cells and later research with many human tumor cell lines indicated that cell death occurred after multiple attempts to undergo mitotic cell division (Tolmach 1961; Elkind, Han and Volz 1963). This mechanism of cell death was called **proliferative** or **mitotic cell death**. Time-lapse cinematography showed most cells to attempt one or two divisions after radiation treatment and only then do they find themselves incapable of unlimited proliferation (Kooi, Stap and Barendsen 1984). Kerr and Seale (1980) described another mechanism of cell death in normal tissues known as **apoptosis**. This is a natural, genetically regulated process for the elimination of specific cells from body tissues that involves the enzymatic breakdown of cellular DNA to fragments of specific size, the involution of the cell membrane and the eventual ingestion of the residual cell debris by phagocytic cells. Since 1980, the number of cancer biology studies each year that invoke apoptosis as the major cancer cell death mechanism after various therapies has risen exponentially. Our study of several human prostate carcinoma cell lines indicated that apoptotic death after radiation treatment was a rare event (Algan et al. 1996). For DU-145 prostate tumor cells, the cell line that showed the largest apoptotic response to radiation, this mechanism of death could account for only a small portion (<5%) of total cell inactivation. For leukemic cells and those of lymphoid origin, apoptotic death appears to be the major mechanism of killing by radiation and several chemotherapeutic agents. This could also be the case for small cell lung cancers (SCLCs) that often shrink rapidly after radiotherapy or even during a course of treatment. But in general, the tumor clonogens in the majority of solid human carcinomas that are treated with radiotherapy will be killed through proliferative death by either the α- or β_0-mechanism.

These details of survival curve generation have been presented to indicate that the determination of precise parameters of radiation inactivation of tumor cells *in vitro* is not a trivial task. One objective of this textbook is to consolidate for research purposes the most representative α- and β_o-inactivation parameters for various human tumors irradiated under all the conditions that might occur *in vivo*. "Radiobiological treatment planning" can then be initiated at every desk by any interested person by inputting the appropriate values into tumor control probability (and possibly normal tissue complication probability) equations. A large number of the radiosensitivity parameters presented in this book were generated in the Chapman laboratories under strict experimental conditions. We have not attempted to obtain "corrected" values of α and β_o from survival data generated in other laboratories (using estimates for repair rates of the β_o-inactivation sublesions) since the dose-rate and temperature during the exposures were not always reported.

Only limited reference is made to studies that were not analyzed in terms of the LQ model. In particular, the research of Mort Elkind and Warren Sinclair that was so important for elucidating several mammalian cell radiobiology mechanisms was not characterized by the LQ equation. We strongly suggest that molecular radiobiologists give consideration to these two independent mechanisms of radiation killing and attempt to relate their research to the LQ model. In this way, novel molecular findings would have a greater impact upon the treatment planning for specific cancers. As improved values of tumor cell parameters important for tumor control probability modeling are determined, these should be rapidly made available to cancer treatment planners.

It is apparent that most survival curves reported in the literature were generated with cell populations of heterogeneous sensitivity at temperatures that did not eliminate the repair of sublethal damages during the radiation exposures. Consequently, their values of β-inactivation will be less than maximal (β_o) and as a consequence their values of α will be larger than the true values.

③ Intrinsic Radiosensitivity of Proliferating and Quiescent Cells

In this chapter the intrinsic radiosensitivity of mammalian cells in different states of proliferation is presented in terms of their α- and β_o-inactivation rates. The cross sections for cell inactivation by these two mechanisms vary significantly and systematically throughout the cell growth cycle. Mitotic cells are uniquely radiosensitive by virtue of very large α-inactivation cross sections, which are similar to those of most repair-deficient cell lines. In G_1-, S- and G_2-phases of the cell cycle, the majority of variation in intrinsic radiosensitivity results from β_o-inactivation, which is highest in the early G_1-phase and lowest in the late G_2-phase. When genome multiplicity is accounted for, α-inactivation is relatively constant throughout the cell cycle but varies significantly between different cell lines. Consequently, when asynchronous populations of mammalian cells are investigated in radiobiology experiments, their survival curves will be complex combinations of the several different radiosensitivities exhibited by their subpopulations. Non-proliferating, quiescent cells, which contain a diploid complement of DNA (similar to G_1-phase cells), are more resistant to radiation killing by virtue of lower values of α-inactivation than proliferating diploid cells. Consequently, cells in tumors and normal tissues that are not proliferating (quiescent) at the time of radiation are likely to be radioresistant relative to proliferating cells.

The clonogens in human tumors can be in an actively dividing (proliferative) or a resting (quiescent) state. Proliferating cells, in general, are more sensitive to ionizing radiation than those that are not dividing (Chapman, Todd and Sturrock 1970; Hahn and Little 1972). In the 1960s, Terasima and Tolmach (1961) and Sinclair and Morton (1963) reported on the radiosensitivities of HeLa and Chinese hamster cells, respectively, as they progressed through their division cycles. Although the doubling times of these two cell lines were considerably different, their data showed that cells are ultrasensitive at the time of mitosis, are relatively radiosensitive in G_1- and late G_2-phases and are most radioresistant in the late S-phase. Thus, it was apparent that cells irradiated as asynchronous populations would express some intermediate radiosensitivity between the most sensitive and the most resistant cells. And each fraction of radiotherapy dose would selectively kill the mitotic and other radiosensitive cells with the most resistant cells surviving.

The division cycle of mammalian cells is defined by four different phases that can be delineated by microscopic examination of their chromosomes and measurements of DNA synthesis (Howard and Pelc 1953). Mitosis is the final phase of the division process whereby cells with a tetraploid complement of DNA divide into two daughter cells, each with a diploid complement of DNA. This is the process by which one fertilized egg eventually produces about 100 trillion cells in an adult human and by which tumors proliferate to produce their lethal effects. At the time of cell division, the chromatin in cells condenses into structures that can be visualized as X- and Y-shaped structures (chromosomes) by light microscope procedures. Human cells at mitosis contain 23 pairs (46 in all) of chromosomes of variable size that contain four copies of their DNA (they are tetraploid). After cell division (cytokinesis), each daughter cell will contain a diploid complement of DNA that must then undergo another round of replication prior to the next cell division. The G_1-phase and G_2-phase designate "gaps" in the growth cycle between the completion of cell division and the onset of DNA synthesis and the end of DNA synthesis and the onset of mitosis, respectively. Many cell lines derived from human tumors exhibit doubling times during exponential growth *in vitro* of 20–30 hr. The time to complete DNA synthesis is ~8 hr, to initiate and complete mitosis is ~1 hr, to accomplish G_2-phase is 3–5 hr and the remainder of the cycle time is the G_1-phase, the most variable of the different growth phases (6–12 hr). Figure 3.1 shows a schematic of this cell division process. It was devised as a tool to teach radiation oncology residents about the cell growth cycle and the effects of radiation on cells in its different phases. It was patterned after a monopoly game whose rules are briefly explained in the figure legend and it certainly whetted the residents' appetites to learn about this biological process that is so important for understanding tumor growth and radiotherapy procedures. The bank notes of the game (dealt to all players) were required to pay the banker in order to progress the number of spaces (determined by a dice throw) around the cycle and were in units of the growth-regulating molecules (CDKs/cyclans) that are unique for progression through specific phases. In addition to mitosis, G_1-, DNA synthesis (S) and G_2-phases, the figure shows a G_0-phase (quiescent phase) and critical DNA assessing G_1- and G_2-phase checkpoints (Dasika et al. 1999; Iliakis et al. 2003; Xu and Kastan 2004). When players landed on any

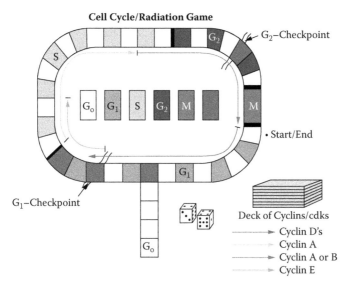

Figure 3.1. A teaching tool devised at the Fox Chase Cancer Center for radiation oncology residents to learn the details of the tumor cell growth cycle. It was patterned after the "monopoly" board game where each resident attempted to progress through the cell cycle and be first to divide into two new cells. One dice throw determined the number of spaces to progress through the G_1-, S- and G_2-phases and mitosis of the cell cycle. The bank manager was paid in units of cyclin/CDKs (dealt to participants at the beginning of the game) required for progression through the various stages and, when one's game piece landed on a colored square, a radiation risk associated with that part of the cell cycle was assessed. The G_0-phase (quiescence) was often the penalty to miss a few turns at the dice as were the G_1- and G_2-checkpoints.

colored square, they picked up a corresponding risk (radiation exposure) card and executed its command (return to origin, miss next three turns for repair of sublethal damage, etc). The first player to reach the first mitosis and divide into two daughter cells was the winner. The cell cycle checkpoints will be discussed in greater detail in Chapter 6.

Mammalian cells have been "synchronized" by several procedures to various degrees of purity for laboratory investigations. These include the selection of cells by their volume (early G_1-phase cells are one-half the volume of a late G_2-phase or mitotic cells) with centrifugation/elutriation techniques (Kauffman et al. 1990). This is a relatively labor-intensive procedure, is somewhat "rough" on the cells and yields only modest amounts of cells for investigation. Others have used chemicals to block cell cycle progression in a pre-DNA synthesis phase and, upon release of this block, follow the surviving cells as they progress into DNA synthesis and beyond (Tobey 1973; Tobey, Valdez and Crissman 1988). The cells that survive this procedure are not necessarily equivalent to late G_1-phase cells since their protein synthesis was not inhibited to the same extent as was their DNA synthesis.

Another procedure utilized the toxic property of radiolabeled DNA precursors to "suicide" cells into which they had been incorporated (Whitmore and Gulyas 1966). It is difficult with any of these procedures to produce "tightly synchronized" populations of cells for subsequent investigation.

The "gold standard" for producing synchronized populations of mammalian cells for both radiation and drug studies is the "mitotic shake-off" procedure. This was the technique used by Terasima and Tolmach (1961, 1963) and Sinclair and Morton (1963). It is based upon the observation that some cells in interphase, growing as monolayers on surfaces, take the shape of a "fried egg" but "round up" at mitosis as they undergo cell division. When a tangential force (a shake) is applied to a culture flask of asynchronously growing cells, the rounded mitotic cells can be selectively released. Selected (rounded) cells are then isolated by centrifugation and placed in culture flasks for completion of division (cytokinesis) and further growth. The Chapman laboratory typically utilized several 500 mL flasks of asynchronously growing cells with Ca^{++}-free media (calcium is required for cell adhesion) and shook them on a reciprocating shaker table at a fixed speed and stroke length. When this procedure was repeated each hour with cells incubated at 37°C in complete growth media between the shakes, populations of cells with mitotic figures of 65%–97% for various cell lines could be produced after the third or fourth shake (Gillespie et al. 1975; Biade, Stobbe and Chapman 1997). It should be noted that this procedure does not work with every human tumor cell line and that the optimal shaker reciprocation speed, stroke length, selection medium and time of intershake incubation are different for the different cell lines. Nevertheless, these techniques produced populations of relatively pure mitotic cells that were capable of undergoing subsequent cell cycle progression without any significant delay. The data shown in Figure 3.2 (from Gillespie et al. 1975) were generated with Chinese hamster fibroblast cells selected in this manner and incubated for the time in hours indicated on each survival curve. Microscopic examination of the selected populations used in this study showed purities of 80%–87% mitotic figures. The average doubling times of these populations were ~11 hr. In some studies, DNA synthesis assays were performed on portions of the cells to determine the approximate timing of the DNA synthetic phase within the cell cycle. The lines through the survival data are best fits to the linear-quadratic (LQ) equation and pass through the majority of survival data points.

Figure 3.3 shows α- and β_0-inactivation parameters for these cells as they traverse the first cell cycle after mitotic selection (Gillespie et al. 1975). The graph contains data from four independent experiments performed on different days for mitotic cells that were subsequently incubated attached to petri dishes (two experiments) or growing in suspension culture (two experiments). This Chinese hamster cell line (V79-379-A) was obtained from the Mort Elkind laboratory (via the Gray Lab) and conditioned to grow in suspension culture (low Ca^{++} media) at AECL (The cell line was shared with the Lloyd Skarsgard laboratory and is described in some of their publications.). The cell cycle times for these experiments were measured to be 10.2–11.6 hr with G_1-phase, S-phase, G_2-phase and mitotic times of ~2.9, ~5.7, ~2.2 and ~0.5 hr, respectively.

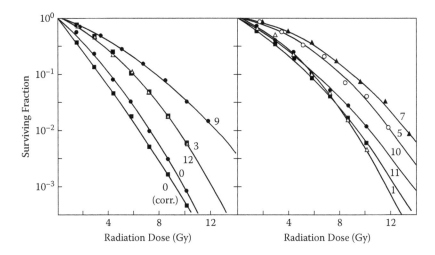

Figure 3.2. Survival curves for Chinese hamster cells selected by mitotic shake-off procedures and irradiated immediately or at various times (indicated in hours on the various curves) after growth through the first cell cycle. The initial mitotic index of this population was 87% and their doubling time as a suspension culture was about 11 hr. (Adapted and reproduced with permission from Gillespie, C. J. et al. 1975. *Radiation Research* 64:353–364.)

It is evident that both α and β_o vary significantly and differently during the cell cycle. The relative radiosensitivities of these hamster cells were similar to those described in a previous report (Sinclair and Morton 1963). Mitotic cells were the most radiosensitive and late S-phase cells were the most radioresistant. It is also apparent that by the end of the first cell cycle significant degradation of the synchronized state of the populations had occurred, probably resulting from an intrinsic distribution in cell cycle times. The radiation parameters were subjected to "goodness of fit" analyses to both the single-hit, multitarget (SHMT) and the LQ equations (Gillespie et al. 1975).

The LQ equation was found to describe the majority of the interphase survival curves accurately, whereas the SHMT equation never did. Since two parameters were extracted from these seven-point survival curves, the normalized sum of squares of the deviations was compared with the 95% χ^2 value for 5 degrees of freedom whose numerical value was 11.07. This analysis did not include the various errors that would undoubtedly arise from the cell handling procedures.

As cells progress through the mitotic division cycle, their diploid DNA content is duplicated by the end of the DNA-synthetic phase. As a result, late S-phase, G_2-phase and mitotic cells will contain a tetraploid complement of chromatin (the radiation target) compared to diploid G_1-phase cells. When the survival data of Figure 3.3 (those generated with cells in suspension culture) were corrected for target multiplicity (diploid = 1 and tetraploid = 2) prior to fitting by the LQ equation, the resulting α- and β_o-inactivation rates shown in Figure 3.4 were obtained

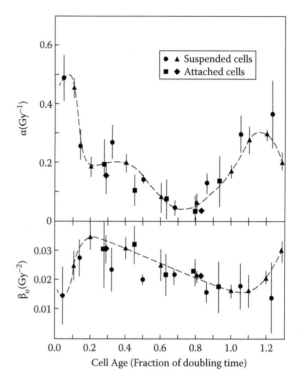

Figure 3.3. The best-fit values of α- and β_o-parameters of the LQ equation throughout the first cell cycle. Data are from four independent experiments for cells incubated attached to plastic dishes (two studies) and growing in slowly stirred suspensions (two studies). The error bars represent the 68% confidence contour for joint variation of the parameters. (Adapted and reproduced with permission from Gillespie, C. J. et al. 1975. *Radiation Research* 64:353–364.)

(Gillespie et al. 1975). These data show that there are two different α-inactivation states for dividing cells: a uniquely high value for cells at mitosis and another smaller value for cells in interphase. On the other hand, β_o-inactivation is seen to be maximal for early G_1-phase cells and decreases to a minimal value in the late G_2-phase, just prior to mitosis. The possible relevance of this finding for identifying radiation targets within cells will be addressed in Chapter 6. This research confirmed that the LQ equation describes the radiation inactivation of dividing mammalian cells extremely well when populations are homogeneous with regard to cell cycle position and irradiated at low temperature to minimize sublethal damage repair.

At the FCCC, radiation studies with synchronized cells were repeated with three human tumor cell lines: HT-29 (colon carcinoma), OVCAR10 (ovarian carcinoma) and A2780 (ovarian carcinoma) cells (Biade et al. 1997). Their cell cycle times were quite similar (19–23 hr) but significantly longer

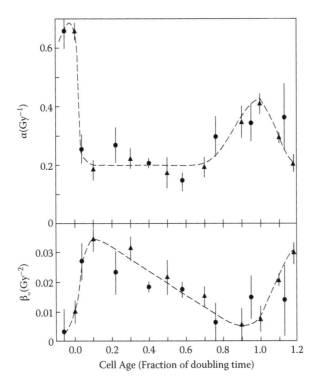

Figure 3.4. α- and β_0-inactivation parameters for the survival data of Figure 3.3 (the two experiments were for cells grown in suspension culture) when corrected for genome multiplicity. These parameters are cross sections for radiation killing per diploid genome. (Adapted and reproduced with permission from Gillespie, C. J. et al. 1975. *Radiation Research* 64:353–364.)

than those of Chinese hamster cells and their intrinsic radiosensitivities (for asynchronous populations) were significantly different. Asynchronous HT-29, OVCAR10 and A2780 cells exhibited α-inactivation values of 0.03, 0.16 and 0.47 Gy^{-1}, respectively, with β_0-inactivation parameters that were almost identical (0.053–0.073 Gy^{-2}). These survival curves are shown in Figure 3.5 along with those for the same cells at mitosis. The purity of mitotic populations that was achieved by our shake-off procedures was ~95%, ~80% and ~65% for HT-29, OVCAR10 and A2780 cells, respectively. Consequently, the LQ equation was fitted to survival data down to only 10%, 20% and 35% for these cells, respectively. Figure 3.6 shows the averages of best-fit values of α (Gy^{-1}) and $\sqrt{\beta_0}$ (Gy^{-1}) each from four experiments (from survival curves of seven to nine dose points) plotted versus the time to the next mitosis after the mitotic cells are plated. Alpha inactivation is relatively constant throughout interphase and the order of radiosensitivity from most sensitive to most resistant is A2780, OVCAR10 and HT-29.

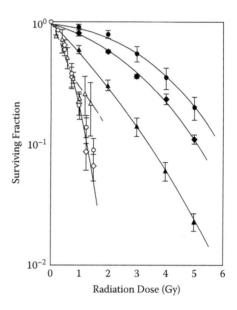

Figure 3.5. Survival curves for asynchronous (solid symbols) and mitotic (open symbols) populations of A2780 (triangles), OVCAR10 (diamonds) and HT-29 (circles) fitted by the LQ equation. (Adapted and reproduced with permission from Chapman, J. D. 2003. *International Journal of Radiation Biology* 79:71–81.)

The values of $\sqrt{\beta_o}$-inactivation are highest in the G_1-phase and lowest in the G_2-phase and are similar for the different cell lines. These studies were consistent with our previous research with synchronized Chinese hamster cells and indicated that the large difference in the intrinsic radiosensitivity of asynchronous tumor cells resulted from significant differences in their α-inactivation parameters in the interphase. The data in Figure 3.5 also show that at mitosis these tumor cells exhibit equal and maximal radiosensitivity, expressed mainly as high values of α-inactivation. This effect will be discussed in more detail in Chapter 6 when ionizing events and molecular targets are addressed. To our knowledge there are no other reports in the literature of α- and β_o-inactivation parameters for mammalian and/or human tumor cells throughout their complete growth cycles. There have been numerous investigations with synchronized cells that address the importance of cell cycle checkpoints, the effect of drugs and several molecular factors on radiation responsiveness, but none with the intent of delineating quantitative values for the α- and β_o-inactivation mechanisms.

Cells in most body tissues are usually not progressing through the cell cycle and contain one diploid complement of DNA. This state has been designated as the quiescent-, stationary- or G_0-phase (see Figure 3.1). The cells in muscle, liver, kidney, brain and other tissues are mainly in this state, are differentiated and metabolize efficiently to produce the specific enzymes and unique molecules associated with their organ function. The epithelia of the gut and bone

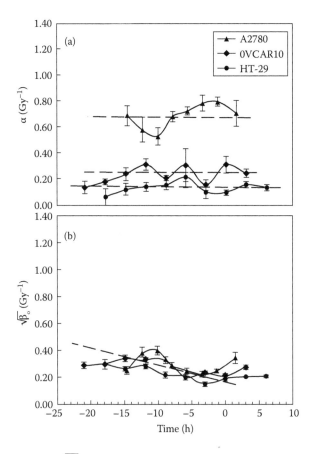

Figure 3.6. α and $\sqrt{\beta_o}$-inactivation parameters for A2780, OVCAR10, and HT-29 human tumor cells throughout the first cell cycle after mitotic cell selection. Since their cell cycle times were different, the parameters are plotted as times prior to the first cell division (0 hr). The initial mitotic indices were ~65%, ~80% and ~95% for A2780, OVCAR10 and HT-29 cells, respectively. (Reprinted with permission from Biade, S. et al. 1997. *Radiation Research* 147:416–421.)

marrow cells will have a larger proportion of their cells in a proliferative state. When cells are irradiated in the G_0-phase and then stimulated to proliferate, they exhibit a lower radiosensitivity than diploid cells in the G_1-phase (Hahn et al. 1968; also see Figure 4.3 in Chapter 4). This could result from longer times available for DNA repair prior to re-entering the proliferation cycle for colony assays or from different conformations of their chromatin. In fact, the irradiation of proliferating cells induces a growth delay that is proportional to the total dose received (~1–2 hr/Gy), which should also be advantageous for promoting DNA repair (Elkind et al. 1963). Many cells when cultured *in vitro* stop dividing when they become contact inhibited or after they have depleted their nutrient medium (Tobey 1973). When such cells (stationary phase) are dispersed into fresh media in culture flasks at reduced density, they can recover their ability to undergo mitotic cell division (Chapman, Todd and Sturrock 1970). The times required to re-enter the division cycle increased as the time held in the plateau phase increased. As well, the volumes (corresponding to mainly protein content) of plateau phase cells were significantly smaller than cells in the G_1-phase and part of the recovery time into the division cycle involved the resynthesis of depleted proteins. Consequently, tumor cells (in quiescent phase) will exhibit a greater radioresistance than those proliferating through the G_1-phase (Chapman and Gillespie 1981). This factor could be important for modeling the treatment response of solid tumors since clonogens can be in proliferating or quiescent states.

If the intrinsic radiosensitivities of proliferating and quiescent tumor cells are significantly different, can the LQ equation be used to describe the radiation response of heterogeneous mixtures of cells? In the study described previously for Chinese hamster cells, an asynchronous population was "modeled" with 10 different subgroups of cells throughout the growth cycle (Gillespie et al. 1975). Cells in each compartment (there would be twice as many newly divided G_1-phase cells as late G_2-phase and mitotic cells) were inactivated by the α- and $β_o$-values that characterized each specific phase (see Figure 3.3). The generated survival curve for this hypothetical population (see Figure 3.7) could be satisfactorily fitted by the LQ equation. The inactivation parameters obtained do not agree exactly with the survival curve of an asynchronous population of the same cells, but cell-cycle analysis was not performed to verify that the two populations were identical

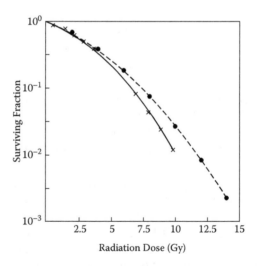

Figure 3.7. The survival curve for an asynchronously growing Chinese hamster cell population (•) mathematically generated with data from Figure 3.3 compared to that of an asynchronous "real" population (x) from a growth flask on a given day. (Adapted and reproduced with permission from Gillespie, C. J. et al. 1975. *Radiation Research* 64:353–364.)

among the various cell cycle positions. We recommend that bars be placed over these $\bar{\alpha}$ and $\bar{\beta}_o$ parameters derived from cell populations with mixtures of cells of different radiosensitivities to distinguish them from those derived from relatively homogeneous populations.

When the LQ equation is fitted to survival data generated with cells of known heterogeneous radiosensitivity, the following equation should be utilized:

$$SF = \exp\left(-\bar{\alpha}D - \bar{\beta}_o D^2\right) = \sum n_x \left(e^{-\alpha(x)D - \beta_o(x)D^2}\right) \tag{3.1}$$

where

SF pertains to the whole population

$\alpha(x)$ and $\beta_o(x)$ are the inactivation parameters for the different subpopulations

n_x = the fraction of the cell population whose radiation response are characterized by $\alpha(x)$ and $\beta_o(x)$

Conceptually, Equation (3.1) should be utilized for characterizing populations of asynchronously growing cells, for mixtures of aerobic and hypoxic cells and for mixtures of proliferating and quiescent cells—all conditions that probably prevail in solid human tumors. However, it is unlikely that such detailed tumor biology and radiobiology data will ever be available for tumor control probability (TCP) modeling. In practice, best-fit values of $\bar{\alpha}$ and $\bar{\beta}_o$ can be obtained for most human tumor cell lines, but exactly how these parameters relate to the radiation–inactivation mechanisms of the various cell subpopulations will probably never be realized.

With regard to clinical outcome data, cell killing by fractionated radiotherapy doses is the required parameter. A typical daily fraction of a 1.8–3 Gy dose will eliminate at least 99% of all tumor cells in mitosis at the time of treatment. For the most radioresistant cells in the late S-phase, only 10%–20% will be killed by the same dose and cells that are quiescent (not undergoing division) will be even more resistant. This differential killing will produce a different distribution in mixed radiosensitivities $\left(\bar{\alpha}\ \text{and}\ \bar{\beta}_o\right)$ from the preirradiation one, so, by the time of the next treatment on day 2, the tumor cell population to be irradiated could be different from that treated on day 1. And this will continue throughout the course of fractionated radiotherapy. This process is known as one of the Rs of radiotherapy: ***redistribution***; it has generally been assumed that over several weeks of treatment, the clonogen radiosensitivities would "average out." For daily fraction sizes of 1.8–3 Gy, where the cell-cycle variation in radiation sensitivities is small and where cell killing by $\bar{\alpha}$-inactivation dominates, this assumption is probably justified. It would certainly make for laborious clinical research to attempt to measure the various compartments of proliferating and quiescent cells in an individual human tumor throughout the course of a specific cancer treatment. As a note of caution; with larger dose fractions, where cell-cycle variations are much larger, or when interfraction times are shorter (than, say, 6 hr), a "perfect" redistribution of tumor clonogens between the dose fractions should not be assumed.

One "take-away" message from this chapter is that the clonogens of consequence in individual human tumors will probably be in various phases of proliferation during treatment protocols and each cell-cycle phase will have a unique intrinsic radiosensitivity defined by the LQ equation. Consequently, $\bar{\alpha}$ and $\bar{\beta}_0$ parameters that describe their radiation killing will be some complex average over several subpopulations of cells. And these could vary from day to day during fractionated treatments, particularly with larger fraction sizes. Such variations in intrinsic radiosensitivity of the clonogens of human tumors could be important for tumor-response modeling but are never routinely measured for individual patients.

④ Effects of Ionization Density and Volume

This chapter describes the effects of ionization density and ionization volume on cell inactivation by the α- and β_o-inactivation mechanisms of the linear-quadratic (LQ) model. Radiation events of very high ionization densities (for example, charged-particle Bragg-peak radiation and Auger cascades) produce cell killing by mainly single-hit, non-reparable lesions, which are not readily modified by chemical environments. High-energy charged-particle beams designed for cancer treatment produce cell killing by both LQ mechanisms in proportions defined by their average linear energy transfer (LET). Sparsely ionizing (low-LET) radiations produce tumor cell killing by both LQ mechanisms and their single-hit killing correlates with the dose deposited by electron track-ends. Since the majority of radiotherapy delivered today is by photon beams from linear accelerators that produce mainly sparse ionizations, tumor cell killing and tumor treatment response will result from both α- and β_o-inactivation mechanisms and in a proportion that can be predicted from the size of the dose fraction and the intrinsic radiosensitivity of the clonogens of each specific tumor. For proton and carbon-ion radiations, their superior dose-depth distributions will allow for significantly higher doses to planned treatment volumes (PTVs) and increased biological effectiveness might be exploited for carbon ions.

In this and the next chapter, the impact of several physical factors involved in cancer treatment protocols will be described. At the time of writing, it would seem that the "radio" component of radiobiology has gone out of fashion when it comes to funding research into cell-killing mechanisms. The "biology" component of the science is now dominated by molecular biology approaches. Nevertheless, it is the physics of radiation deposition (its microdosimetry and nanodosimetry) within cells that informs about cell killing by both α- and β_o-mechanisms of the LQ model, and "target theory" should be resuscitated.

The vast majority of radiotherapy today is delivered by high-energy photon sources. Some is delivered by the decays of radioisotopes that are implanted directly into solid tumors. Over the past 30 years, there have been several proton sources developed for dedicated clinical applications and a few carbon-ion sources. The practice of radiation oncology can now exploit an array of novel radiation sources that are capable of delivering therapeutic dose much more precisely to PTVs.

4.1   Ionizations along Charged-Particle Tracks

Early radiobiology studies with cyclotron-generated beams of charged particles showed that the passage of one particle through the nucleus of a mammalian cell produced cell killing. Barendsen et al. (1963) and Barendsen (1968, 1990) investigated α-particles and deuterons and Todd (1967, 1974) investigated heavier charged particles. The inactivation cross sections derived from these studies were 40–100 μm^2 for cells whose nuclear cross sections were 90–100 μm^2. Cell killing by these particles (Bragg peak irradiation) was by single-hit inactivation (α) that showed no potential for repair. The radiation chemistry of these one-particle inactivation events was qualitatively different from multiple-particle inactivation, in that they exhibited no radiosensitization by oxygen or radioprotection by sulfhydryl compounds (Barendsen and Walter 1964). These studies established that the most efficient ionization density for killing mammalian cells was at a LET of 100–150 keV/μm, with LET being defined as the energy deposited in electron volts along the track of a moving charged particle in micrometers.

Other studies performed at the M. D. Anderson Hospital and Tumor Institute by Cole et al. (1974), Cole et al. (1975) and Tobleman and Cole (1974) utilized electron beams of low energy that could penetrate only 1/10 to 1/3 into mammalian cells growing as monolayers on a surface. In this case, radiation killing was also described by single-hit kinetics (α), showed no potential for repair and exhibited no radiosensitizing effect by oxygen. These studies indicated that even stopping electrons (track-ends) could produce cell killing with the characteristics of the heavier charged particles. This effect was referred to as the high-LET component of low-LET radiation.

The early studies of charged-particle radiobiology utilized relatively low energy particle beams that would have little or no utility in conventional cancer treatment. They were accomplished with mammalian cells irradiated as monolayers on glass or plastic cover slips that were placed at right angles to the particle beam for exposures. Similar techniques were used in the low-energy electrons studies. It is obvious that cells in tumors do not grow as monolayers on supporting surfaces (with a fried egg configuration), but this technique was well suited for investigating the radiation effects of relatively narrow LET distributions. The monolayer irradiation procedure was not optimal for investigating synchronized cells of homogeneous radiosensitivity or for measuring the effects of various chemical environments on radiation response.

For these reasons, the glass chamber described in Chapter 2 was constructed at LBL for irradiating mammalian cells in cubic centimeter volumes of slowly stirred suspensions where they would be approximately spherical (a configuration closer to that of cells *in vivo*) (see Figure 2.2 in Chapter 2). This technique facilitated the investigation of both asynchronous and synchronized populations when exposed in various chemical environments at different temperatures. The trade-off that was made by using that irradiation chamber was for improved biological homogeneity at the expense of broader distributions of stopping particle LETs. The chamber has a depth of 1.5 cm and the resultant survival curves were stirred averages of cell killing by a mixture of particle track segments (see Figure 4.1). The speed of cell stirring by a magnetic bar was high enough to guarantee that cells would be in every position in the chamber volume over the exposure times of about 1–2 min. For the charged-particle beams produced at the BEVALAC accelerator in 1975 and 1976, the distribution of LETs sampled by this chamber was much broader than that for cells irradiated as monolayers (Blakely et al. 1979). For cancer therapy, it is the killing effectiveness of the radiation deposited in the PTV that is of consequence.

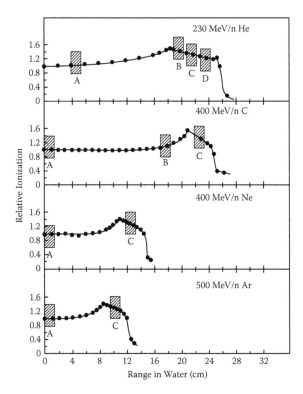

Figure 4.1. The depth-dose profiles of 230 MeV/AMU He- and 400 Mev/AMU C-, Ne- and Ar-ions produced at the BEVALAC indicating the positions in the particle tracks where the irradiation chamber (Figure 2.2, Chapter 2) was positioned. (Reproduced with permission from Chapman, J. D. et al. 1977. *International Journal of Radiation Oncology Biology Physics* 3:97–102.)

The averaging of cell killing in stirred volumes of a few cubic centimeters should be equivalent to that obtained by mathematically adding cell killing in multiple cell monolayers of 10 µm depth. The optimization of both the physical and biological factors that can modulate cell radiosensitivity in the same experiment can be extremely difficult and sometimes impossible.

Chapman's research at AECL had defined various chemical pathways involved in cell killing by some low-LET radiations (see Chapter 6) and his research at LBL was able to extend these studies to high-LET radiations. That research had emphasized the role of direct effects and those of water free radicals in the radiation killing of mammalian cells *in vitro* and the effect of various chemical modifiers on those killing mechanisms (Chapman et al. 1973). The BEVALAC accelerator (a "marriage" of the super HILAC and Bevatron accelerators at LBL) research program had just begun in 1974, and one study by the Tobias group involved the characterization of these high-energy beams for cancer therapy. The beams available for study at that time were carbon, neon and argon ions from the BEVALAC and helium ions from the 184-inch cyclotron. Dr. Eleanor Blakely was coordinating the *in vitro* cell radiobiology investigations for the Tobias group using human kidney T cells irradiated as monolayer cultures on glass cover slips (later in specially designed pertri dishes), and she became a major collaborator in Chapman's research. The cover slip experiments employed a chamber filled with sterile tissue culture media affectionately known as "the submarine chamber" (it looked like one). With this technique, cells could be exposed to a relatively narrow distribution of LETs at various positions along the stopping particle tracks whose initial energies were large enough to penetrate up to 30 cm in water. It was a somewhat cumbersome procedure involving the plating of cells on cover slips prior to irradiation, removing them from the cover slips after irradiation by an enzyme procedure and plating known numbers (involving a Coulter count for each dose point) into petri dishes for colony-forming assays (Blakely et al. 1979). This cover slip technique was replaced later by a special irradiation chamber for many of their studies (Blakely et al. 1984).

Another component of that program project was the characterization of cell killing in rodent tumors (growing on the thighs of mice) by the same charged-particle beams. That study was supervised by Professor Ted Phillips, of the UC Medical Center in San Francisco, who transported animals across the Bay Bridge whenever he was alerted about potential beam time. Cells in rodent tumors of ~0.5 cm³ volume would necessarily be exposed to broader distributions of LET relative to those experienced by the cell monolayers. The "Canada" radiation chamber (Figure 2.2 in Chapter 2) exploited an intermediate strategy—the averaging (by stirring cells) of radiation effectiveness over a 1.5 cm length of stopping particle tracks. In hindsight, that was a wise decision since it expedited the investigation of the He-, C-, Ne- and Ar-beams over spatial dimensions important for mouse and human tumors. And these studies could be readily performed with both asynchronous and synchronized cells for numerous conditions of defined chemical modification and at several temperatures. The reliability of the BEVALAC in its early years of operation was spotty at best and biology usually got the late evening to early morning time slots. Consequently, several populations of mammalian cells that had been synchronized for investigation were never exposed, but 156 survival curves, usually consisting of eight different dose points, were produced for analysis by the LQ model. Stationary-phase cells (of greater biological homogeneity than asynchronously growing cells; see Chapter 3) were often the fall-back population of cells that were investigated.

Figure 4.2 shows values of α and $\sqrt{\beta_o}$ obtained for stationary-phase Chinese hamster cells irradiated at various positions along the stopping paths of He-, C-, Ne- and Ar-ions whose initial energies were 230, 400, 400 and 400 MeV/n, respectively (Chapman et al. 1977). Cell killing was measured in both the plateau and the spread-peak portions of the different beams (see Figure 4.1) and cell-killing parameters were plotted against the average LET (keV/μm) (computed by Dr. Stan Curtis), the LET at which 50% of the dose was deposited. It is obvious that the increased biological effectiveness of these particle beams results mainly from an increase in the α-inactivation mechanism and reaches a peak value at 100–150 keV/μm. The inactivation parameters of similar cells exposed to 220 kVp x-rays are also shown (the data points at 4 keV/μm). The peak in biological effectiveness agreed with the optimal LET for single-hit killing that had been reported with lower energy particle beams (Todd 1967) and also identified significant additional killing by the double-hit mechanism, presumably resulting from low-LET components of particle track penumbras. For stationary phase Chinese hamster cells, α-inactivation increased ~10-fold between average

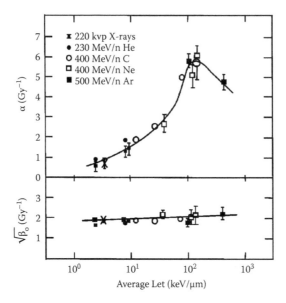

Figure 4.2. Best-fit values of α and $\sqrt{\beta_o}$ parameters (±SD) obtained from seven- to nine-point survival curves for stationary-phase Chinese hamster cells irradiated at various positions of depth dose for various charged particles plotted versus their average LET. (Adapted and reproduced with permission from Chapman, J. D. et al. 1977. *International Journal of Radiation Oncology Biology Physics* 3:97–102.)

LET values of 2 and 100 keV/µm while the $\sqrt{\beta_o}$-inactivation increased by less than 20% over the same range. The low-LET radiation in the energetic tracks and their penumbras produces a background of energy deposition that is no more effective in killing cells than the majority of energy deposited by x-rays. In clinical treatment protocols, the maximal killing effectiveness of Bragg peak radiation of protons and carbon ions will always be diluted by their lower LET components. Consequently, charged-particle radiation with enough energy to penetrate 20–30 cm of human tissue will not exhibit all the radiobiological advantages suggested by the early studies performed with low-energy particle beams. It would be preferable if all the radiation deposited in tumor PTVs had the maximal cell-killing effectiveness, but this cannot be achieved with energetic carbon- or neon-ion beams. Now that cell killing has been quantified for a wide spectrum of average LETs with cell lines of widely different photon radiosensitivity, "biological" treatment planning of various PTVs can be performed, voxel by voxel.

Figure 4.3 shows the α-inactivation parameters obtained for stationary-phase, G_1-phase and mitotic populations of Chinese hamster cells. As for stationary phase cells (see Figure 4.2), $\sqrt{\beta_o}$-inactivation values obtained for G_1-phase and mitotic populations increased only slightly over the range of average LET investigated. The α-inactivation parameters for cells in G_0-, G_1- and mitotic phases for radiations of 100–150 keV/µm average LET (that of spread-peak Ne ions) were 5.5–7.5 Gy^{-1}. The large variation in intrinsic radiosensitivity between these cells observed with low-LET radiation is greatly reduced at this

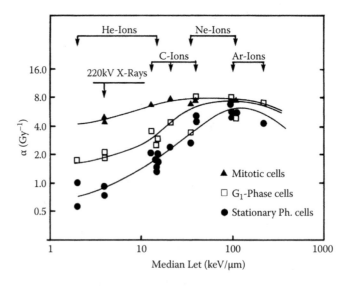

Figure 4.3. Values of α-inactivation for mitotic, G_1-phase and stationary phase Chinese hamster cells irradiated at various positions of depth dose for various charged particles plotted versus median LET. (Adapted and reproduced with permission from Chapman, J. D. and Gillespie, C. J. 1981. *Advances in Radiation Biology* 9:143–198, Academic Press.)

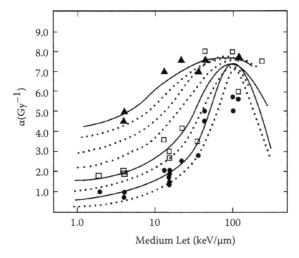

Medium Let (keV/µm)

Figure 4.4. The data of Figure 4.3 with interpolations that could be used for generating tumor treatment plans, voxel by voxel. When the α-inactivation coefficient for any tumor cell line generated with low-LET radiation is known, the expected gain in biological effectiveness for any track segment of various charged-particle beam dose is apparent. The β_o-inactivation parameter can be assumed to be constant (see Figure 4.2).

optimal LET. This indicates that those tumors that are most resistant to photon therapy (see Chapter 7) should benefit the most from treatments with carbon- and neon ions (Figure 4.4).

If confirmed by additional experiments these parameters of cell killing could be incorporated voxel by voxel in the treatment plans for PTVs of human tumors of different pathologies. Particle beams that exhibit the largest increase in α-killing between their transit dose and Bragg peak dose should be optimal for the treatment of photon-resistant cancers. But this cell killing will be in addition to that by β_o-inactivation produced by the low-LET component of the radiation fields, which is surprisingly constant. From these studies, carbon ions were deemed to have the maximal benefit by these criteria (Chapman et al. 1978).

The reader will note that relative biological effectiveness (RBE) has not been employed to describe this increased effect of cell killing by high-LET particles. RBE is defined as

$$RBE = D_s/D_x \qquad (4.1)$$

where D_s and D_x are radiation doses of a standard and novel radiation regimen, respectively, that produce equivalent biological effect.

The description of the data in Figure 4.2 in terms of RBE would only confuse our understanding of basic radiation mechanisms. **Values of RBE depend upon the cell line, the quality of the standard radiation and the dose fraction size.** RBE will increase as the dose fraction size decreases since the majority of the increased effectiveness is by α-killing, which is increasingly more dominant at lower doses (Chapman et al. 1978). The maximal RBE as fraction size tends to zero (the ratio of α-inactivation for spread-peak neon ions versus x-rays) for G_0-phase, G_1-phase and mitotic cell populations is about 10, 4.7 and 1.5, respectively. This wide variation results mainly because of their variable radiosensitivity to 220 kVp x-ray (the standard radiation in this study), which is the numerator in Equation (4.1). Since several different cell lines with different low-LET sources (standard) have been used in various laboratories to characterize biological effectiveness, it should be expected that significant differences in RBE values would be reported for neutrons, protons and other charged particles. The reporting of biological effectiveness in terms of absolute intrinsic radiosensitivity (α and β_o) for one standard cell line would lead to greater clarity. Belli, Campa and Ermolli (1997) have proposed alternative approaches to determining the RBE of clinical proton beams.

The Tobias group at LBL reported the killing effectiveness of various charged-particle track segments on human kidney T-1 cells (Blakely et al. 1979, 1984). This research is a *tour de force* that stands as the basis for much of today's clinical interest in high-LET radiation sources. It characterized particle track segments with much tighter distributions of average LET than were possible with the radiation chamber used to generate the data in Figures 4.2 and 4.3. Their data also confirm that the increased biological effectiveness of these particle beams was mainly due to large increases in the single-hit inactivation parameter, which peaked at an average LET of 140 keV/μm. The survival data were quantified using three different mathematical expressions and the values of α- and $\sqrt{\beta_o}$-parameters obtained for aerobic cells are shown in Figure 4.5. It is apparent that as the average LET increased from 10 to 150 keV/μm, the killing effectiveness by these particles also increased by virtue of a large increase in α-inactivation with a much smaller increase in $\sqrt{\beta_o}$-inactivation. At LETs larger than 100 keV/μm, the different particles take on some unique characteristics, especially neon ions. This research also showed that the oxygen enhancement ratio (OER) for cells irradiated in these various particle track segments decreased from a value of about 3.0 to 1.2, with the lowest values achieved with the heaviest particles (see Figure 4.6). The earlier studies of Barendsen and Walter (1964) and Todd (1967) had measured oxygen enhancement ratios (OERs) of 1.0 at the optimal LETs, but those studies used particle beams whose initial energies were much lower than those produced by the BEVALAC. If the radioresistance of hypoxic cells in human tumors is the major *raison d'etre* for exploiting charged-particle beams of higher LET, the sources developed for clinical use should not be expected to produce equal radiosensitivity for aerobic and hypoxic cells at average energy densities of 100–150 keV/μm. When the OER is averaged over larger volumes of charged-particle dose (in typical PTVs), values of only 1.7 and 1.6 were observed in 4 cm spread peaks of carbon and neon ions, respectively (Chapman et al. 1977; Chapman, Urtasun, et al. 1978). The advantages of improved depth-dose distributions and increased RBE at depth could become the major advantages for improved tumor cell killing in human cancers. And just how these advantages should be predicted from current research is an ongoing study for several physics groups (Frese et al. 2012).

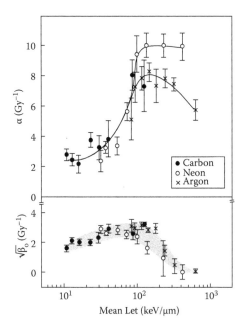

Figure 4.5. The best-fit values of α- and $\sqrt{\beta_o}$-inactivation for human kidney T-1 cells irradiated in air at various track segments of different particle beams versus mean LET. (Adapted and reproduced with permission from Blakely, E. A. et al. 1979. *Radiation Research* 80:122–160.)

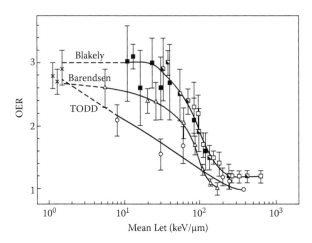

Figure 4.6. The oxygen enhancement ratio measured at the 10% survival level for human kidney T-1 cells irradiated in the various track segments reported in Figure 4.5. The OERs obtained in previous studies by Barendsen and Todd are also shown. (Reproduced with permission from Blakely, E. A. et al. 1979. *Radiation Research* 80:122–160.)

For these LBL data to become useful for clinical treatment planning, the particles' effectiveness in track segments of about 10 μm depths will have to be summed over the track segments that will be deposited in clinical PTVs. And since most of these data were for asynchronous cell populations, the measured $\bar{\alpha}$- and $\bar{\beta}_0$-inactivation parameters are already some complex averages of the different radiosensitivities of cells in the different phases of the cell cycle. Nevertheless, there is good agreement between the killing effectiveness measured for these particle beams using the cell suspension and the cell monolayer techniques. When the BEVALAC was decommissioned in the 1980s, there were attempts to build a medically dedicated carbon-ion accelerator in the San Francisco Bay area. In parallel with that effort, a group of scientists at the University of Alberta initiated the design of a carbon-ion accelerator dedicated to medical research, referred to as MARIA (Medical Accelerator Research Institute in Alberta) (Chapman 1983). Unfortunately, both efforts failed, but the Canadian design study was passed on to scientists at the National Institute of Radiological Sciences in Chiba, Japan, where it served (in part) as the basis for the first clinical carbon-ion radiotherapy center in the world (Kanai et al. 1999). Since that time, three additional carbon-ion radiotherapy facilities have come on stream—two more in Japan and one in Germany. The Hyogo (Japan) facility is capable of producing both proton and carbon-ion beams for comparative clinical studies (Kagawa et al. 2002). Two more carbon-ion treatment centers are under development in Japan. But the current direction of radiotherapy technology developments in Europe and the United States is toward proton beams since they can be produced at a much lower cost and provide physical dose distributions as good as carbon-ion beams (Figure 4.7 from Frese et al. 2012; Elsässer et al. 2010).

When the biological effectiveness measured along a particle track is multiplied by the physical dose, the proton and carbon-ion beams exhibit a 1.5- and 3.0-fold increase in killing potential, respectively, across spread peaks relative to entrance dose (Figure 4.7). These values are dependent upon the intrinsic radiosensitivity at low LET of the cells of interest and the fraction size. The spallation (fragmentation) of heavier charged particles in transit to tumor PTV results in dose beyond the Bragg peak of the primary beam, a phenomenon that does not occur with protons. As a consequence, if a low OER and increase in RBE in the PTV for spread-peak carbon beams are significantly diminished by particle fragmentation, it could be that protons will achieve most of the benefits that can potentially be realized with carbon ions.

4.2   The ABCs of Charged-Particle Radiotherapy

The advantages of charged-particle radiations over conventional linear accelerator (linac) therapy have been described as the ABCs of high-LET radiotherapy. The A refers to **adsorbed dose** and its increase at depths in tissue relative to the entrance dose. Unmodulated Bragg peaks of protons and He-, C- and Ne-ions stopping in water have almost equal amplitudes relative to plateaus. Consequently, equal advantage of improved depth-dose distributions should be expected with

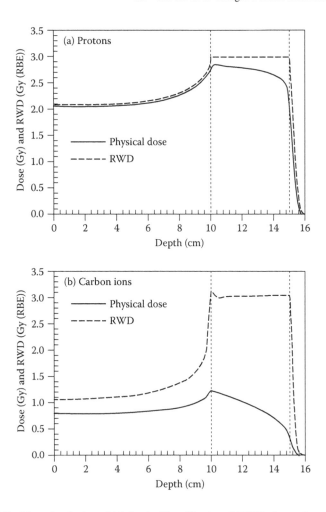

Figure 4.7. The physical and biologically effective (RWD) doses for (a) proton and (b) carbon ion spread peaks whose energies were adjusted to penetrate 15 cm in water. (Reproduced with permission from Frese, M. C. et al. 2012. *International Journal of Radiation Oncology Biology Physics* 83:442–450.)

any of the particle beams characterized in the preceding studies. Of course, the cost to construct patient treatment facilities (especially with *iso*-centric gantries) would increase as the Z-values of the particle of choice increase. But since the physical dose at any depth of tissue relative to the entrance dose is greater than 1.0, very good improvements in relative dose delivery to PTVs should be expected, whether single or multiple ports are compared. Again, the heavier charged particles fragment to lower Z-particles as they penetrate water (tissue), producing dose beyond their Bragg peaks, which must be accounted for in clinical treatment plans (Kraft 2000; Krämer and Scholz 2000). Since protons do not fragment at these energies, they produce no dose beyond their Bragg peaks.

In the ABC description, B refers to the **biological effectiveness** of the physical dose. It was found to increase systematically as the heavier charged particles "slowed down," mainly due to higher values of α-inactivation. The maximal effectiveness of the particles investigated was at an average LET of 100–150 keV/μm (Figures 4.2 and 4.5). The biological effectiveness of specific track segments should be multiplied by the physical dose to obtain their tumor-cell-killing potential. Carbon ions were found to exhibit the largest differential in α-inactivation mechanism between the spread peaks and plateau, suggesting they could be optimal for radiotherapy. Cell killing by 400 MeV/AMU argon ions was already maximal and became less effective as these particles slowed down in water, an effect described as overkill (see Figure 4.2). Proton radiobiology studies have reported increases in biological effectiveness of only 1.0–1.15 at some positions along their stopping tracks in water (Paganetti et al. 2002). If cells are irradiated as monolayers at the exact position of the proton Bragg peak, much larger biological effectiveness will be expressed. But when proton beams are modulated to produce spread peaks over clinical PTVs of several cubic centimeters, their expected biological effectiveness will be close to 1.0. The potential gain for radiotherapy from the differential biological effectiveness of carbon ions along their stopping tracks suggested an advantage for cancer therapy. Clinical research from the existing carbon-ion facilities should establish whether or not this effect can translate into a therapeutic advantage in a cost effective manner.

The C in ABC refers to the **chemical modification** and **cell cycle variations** that can impact biological effectiveness of high-LET radiotherapy. It should be remembered that in the 1970s–1980s, intensive research with hypoxic radiosensitizers had shown selective potentiation of the radiation killing of hypoxic cells in both *in vitro* and animal tumor studies (see Wasserman and Chapman 2004). The involved researchers anticipated that much of the reduced OER observed with charged particle radiations would be clinically achieved with oxygen-mimicking drugs. A major driving force for developing clinical sources of higher LET had been the demonstration of equal radiosensitivities of aerobic and hypoxic cells. Research performed at LBL showed that hypoxic cell radiosensitizers produced increased values of both α- and β_0-inactivation for hypoxic cells irradiated with the charged-particle beams (Chapman, Urtasun, et al. 1978). Several involved researchers were so optimistic about the clinical application of hypoxic radiosensitizers that little thought was given to the possibility that none of them would reach routine use. With regard to differential cell cycle effects (see Chapter 3), particle beams of optimal LET were shown to increase the killing of G_0- and G_1-phase cells to a much greater extent than of mitotic cells (Figure 4.3). Further research indicated that the RBE of charged-particle radiation in AT cells (cells with a high α-inactivation coefficient with x-rays) was much lower than that for photon-resistant cells such as human kidney T-1 cells (Tobias et al. 1984). Cell cycle selective killing and redistribution of surviving clonogens in tumors should be less problematic for fractionated charged particle therapy at their optimal LET. This research strongly suggests that the tumors that will benefit most from radiotherapy with high-LET radiations are those that are maximally resistant to photon radiation—those with low values of $\bar{\alpha}$.

4.3   The Frequency of Electron Track-Ends in Radiation Dose

If particle and electron track-ends are the optimal entities for inactivating tumor cells, are there other means of delivering these high-LET entities to tumor PTVs? At least two other techniques that might be useful in radiotherapy have been investigated. Radioisotopes incorporated into cellular compartments and molecules produce lethal damages when they disintegrate (Marin and Bender 1963a, 1963b). Tritium, by virtue of its low-energy, short-range electrons produced at disintegration, was found to produce effective cell killing, especially when the isotope was incorporated into cellular DNA. Burki et al. (1973) showed that iodine-125 was more effective per disintegration and per unit dose in cell killing than was tritium when it was incorporated into cellular DNA. Some of the energy associated with iodine-125 disintegration is dissipated in localized events composed on average of six electrons of 0.5–34.6 keV—events known as Auger cascades (Burki et al. 1973; Nikjoo, Emfietzoglou and Charlton 2008). When this isotope is selectively incorporated into cellular DNA, cell killing is characterized by single-hit (α) inactivation. This class of radiopharmaceuticals could have therapeutic potential if a sufficient number of radioisotopes could be safely delivered to clonogenic cells in human tumors and allowed to decay locally. This technique was designated a "suicide" technique since it was radioactive decays within individual cells that produced the lethal events.

In laboratory studies, cellular DNA was shown to be the most radiosensitive target susceptible to decays of incorporated isotopes. Precursors of DNA synthesis (^{125}I-Udr) and radiolabeled tumor-specific antibodies that internalize to the cell nucleus have been investigated. In order to accumulate a sufficient number of isotope decays to produce cell killing in tissue culture experiments, labeled cells were stored in frozen media that contained some cryogenic preservative. When enough time had elapsed for sufficient disintegrations, the cells were thawed, diluted and plated in fresh media for colony-forming assays. While these techniques have produced several studies of mechanistic importance, it is obvious that their translation to clinical practice is not an easy task.

Of the many studies, those from two research laboratories are described as potentially relevant for the clinic. Adelstein and Kassis at the Harvard Medical School reported on the radiation killing of mammalian cells by isotope disintegrations within different cell compartments. Their research confirmed that it was the Auger electron cascade occurring in close proximity to cellular DNA that was most effective in producing cell kill (Kassis 2004). That research also showed that a large component of the killing (over 90%) resulted from the indirect effect of hydroxyl radicals generated in the water surrounding the DNA target (Walicka, Adelstein and Kassis 1998a, 1998b). These isotope disintegration events exhibited properties similar to those produced by electron track-ends. The mechanistic aspects of this research will be further discussed in Chapter 6. Additional radioisotopes are now available for radiobiological study, and iodine-123 has a large advantage over iodine-125 by virtue of its shorter decay half-life. Kassis (2008) showed that ^{123}I-labeled DNA precursors could be delivered to transplanted tumors in rodents at levels that produced significant tumor response. Presumably, radionuclide uptake was greater into proliferating tumor cells. These studies will have to assess the detrimental effects of radiation in rapidly proliferating normal tissues (bone marrow and gut epithelia) to determine if a therapeutic gain can be expected in humans. Nevertheless, radioisotopes with shorter half-lives (1–5 days) should produce more rapid killing of tumor clonogens by the suicide technique than tritium or ^{125}I-labeled compounds.

Jensen and colleagues in Chicago devised a potentially novel way for applying the isotope suicide technique to breast cancers (Jensen and DeSombre 1973). Their strategy was to administer ^{123}I-labeled antibodies specific to the estrogen receptors of those patients whose breast cancers expressed this genotype; allow time for them to diffuse to, internalize and "bind" to DNA in the cancer cells; and accumulate isotope decays adjacent to the cellular radiation target. The number of isotopes that could potentially be delivered to individual cells by this procedure was upward of 10^5–10^6 and their decay over a few days had a reasonable probability to cause cell death. The best-case scenario for successful radioimmunotherapy includes (1) that all the clonogenic tumor cells would be selectively and adequately labeled with the radiolabeled agent; (2) that the bulk of the antibody dose that does not bind with the cancer would be rapidly excreted, causing little or no damage to the vasculature and other normal tissues; and (3) that there would be no intertissue (tumor to normal) transfer of the agent. There are several different groups investigating these procedures for different tumors and they have recently been reviewed by Denardo et al. (2002). The jury is still out on whether or not such radioimmunotherapy procedures will have significant clinical benefit. The exploitation of radiolabeled agents to produce local and lethal damage selectively to cancer cells *in vivo* (some current attempts use isotopes that disintegrate by α-particle emission) certainly justifies additional experimentation. But biophysical calculations have made it clear that the number of localized isotopes/cell required for inactivation is large (up to 10^6) and that radiotoxicity in normal tissues exposed after systemic administration of such agents could be limiting. As new cancer cell receptors are identified and additional radioisotopes become available, this field of research could produce important clinical advances over the coming years. Targeted radionuclide therapy is subject to all the limitations of drug therapy based upon the targeting of tumors through antibodies that are usually less tumor specific than anticipated.

When it was postulated from microdosimetry considerations that electron track-ends could produce lethal radiation damage in cellular targets (Chapman 1980, 2003) that was qualitatively distinct from that produced by simple ionizations (see Chapter 6), alternative methods for increasing the frequency of such events in cells were investigated. Chinese hamster cells irradiated as monolayers attached to glass petri dishes had exhibited increased killing relative to cells irradiated in plastic dishes due to increased dose from low-energy, backscattered electrons (Chapman, Sturrock, et al. 1970). (That research also measured the oxygen solubilities in and the oxygen diffusion coefficients from several available industrial plastics and identified TPX plastic as optimal for hypoxia studies with cells in tissue culture—the Permanox® pertri dishes were defined by that research.) The electrons backscattered from glass are stopped in water within 10–100 μm from the scattering surface and could potentially produce inactivating events that are qualitatively distinct (more track-ends) from those received by cells irradiated on plastic. Das and Chopra (1995) had physically measured the backscattered dose from several material surfaces and the Chapman laboratory collaborated in a study to measure the killing efficiency of these radiation fields in monolayer cultures of Chinese hamster cells growing in specially manufactured petri dishes with Mylar bases of 6-μm thickness (Zellmer et al. 1998). The dishes were obtained from the Radiological Research Accelerator Facility at Columbia University (generously provided for these studies by Eric Hall and Steve Marino), where they had been utilized in studies of single particle biological effects. The intensity of backscattered dose was a function of the atomic number (Z) of the scattering material and the quality of the incident radiation (Das and Chopra 1995). With incident 200 kVp x-rays, backscattered dose from high-Z materials such as tin and lead could be 7- to 12-fold greater near the scattering material than that experienced by cells irradiated on plastic surfaces (see Figure 4.8). Asynchronous monolayers of Chinese hamster ovary wild-type (CHO-K1) and a DNA repair deficient mutant (CHO-xrs5) were grown in the Mylar-bottom dishes. The average midline of the nuclei in these cells was measured by electron microscope procedures to be ~3.3 μm from the Mylar base that was 6 μm thick. The radiation dose to the attached cells (direct and scattered) was assumed to be that measured at 9.3 μm from the scattering material. The cross-hatched region in Figure 4.8 indicates the average position of the cell nuclei in relation to the scattering surface from which the backscattered doses would arise. Survival curves were generated with different radiation exposures from 200 kVp x-rays and ^{137}Cs gamma-rays scattered from various materials, and best-fit values of $\bar{\alpha}$- and $\bar{\beta}_0$-inactivation parameters were obtained (Zellmer et al. 1998). It was found that the backscatter dose factors (BSDF) measured by the physical dosimetry (Das and Chopra 1995) could account for the majority of increased killing observed. There was also a suggestion that backscattered dose from 200 kVp x-rays from the highest Z materials produced more α-inactivation than would be expected simply by dose enhancement. This effect was attributed to more stopping electrons in the cell nuclei than experienced when cells were irradiated on plastic surfaces. The BSDF could account for the majority of increased killing measured with CHO-K1 cells, but overestimated the effectiveness of the same fields in CHO-xrs5 cells. This suggests that the molecular targets in these cells associated with α- and β_0-killing could be different (see Chapter 6). Since asynchronous cells were utilized in this study, the data are less amenable to mechanistic interpretation than data from studies with cells of homogeneous radiosensitivity. Nevertheless, it was apparent that radiation dose in the vicinity of these high-Z materials was significantly greater and possibly more effective (higher RBE) than the dose experienced at millimeter distances from the same materials.

This study indicated a potential role for high-Z scattering materials in human tumor radiotherapy if its delivery to human tumors was possible. Consequently, gold microspheres were investigated for producing increased dose in cultured cells and in rodent tumors (Herold et al. 2000). Gold microparticles at a concentration

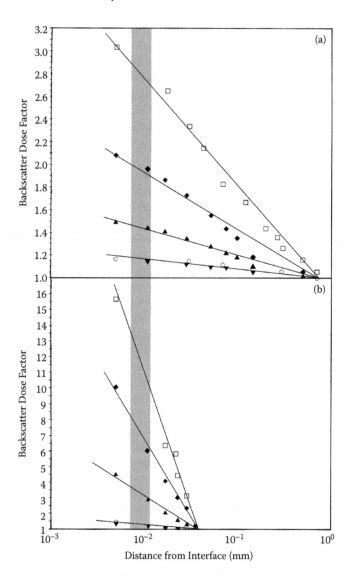

Figure 4.8. The enhanced radiation dose (as a backscattered dose factor, BSDF) at various distances from materials of different density for (a) ^{137}Cs γ-rays and (b) 200 kVp x-rays. The symbols are for glass (inverted solid triangles), aluminum (open circles), copper (solid triangles), tin (solid diamonds) and lead (open squares). The cross-hatched region identifies the position at which Chinese hamster cells (CHO and CHO-xrs5), whose nuclei midlines were 9.3 mm from the scattering surface, were irradiated. (Reproduced with permission from Zellmer, D. L. et al. 1998. *Radiation Research* 150:406–415.)

of 1.0% (in solution) increased the physical dose (as measured by Fricke dosimetry) in slowly stirred suspensions by 50% and increased the radiation inactivation of three different cell lines by the same extent. When these gold particles were injected directly into EMT-6 tumors growing in mice prior to irradiation with 200 kVp x-rays, increased cell killing was measured by *in vivo/in vitro* assays. Since the observed effect was small and the particle delivery was invasive and heterogeneous, it was concluded that this agent would have little or no clinical benefit, especially if linac radiation was employed. Advances in high-Z particle delivery systems to tumors *in vivo* could alter this conclusion, but it is unlikely that radiation oncology will revert to low-energy radiation sources unless a very large benefit can be demonstrated. A new generation of radiobiologists and medical physicists is pursuing this strategy with gold nanoparticles that may be more appropriate for clinical application (McMahon et al. 2008; Lechtman et al. 2011).

In summary, this chapter reviews much of the radiobiology research that identified novel radiation sources and dose-delivery procedures that could produce more effective treatments of human cancers. With the improved definitions of PTVs available today by computerized tomography (CT), positron emission tomography (PET) and magnetic resonance (MR) imaging, the physical dose delivery to these volumes by intensity modulated radiation therapy (IMRT), gamma-knife, tomotherapy, proton and carbon-ion techniques should be greatly improved. In addition to optimal physical dose distributions, it is time to give more attention to radiobiology parameters that might be exploited for improved cancer treatments.

5 Impact of Fraction Size, Dose-Rate, Temperature and Overall Treatment Time on Tumor Cell Response

Dose fraction size, dose-rate and overall treatment time will impact strongly on tumor cell killing and tumor response outcomes. Cell temperature at the time of cancer treatment will be near 37°C but most laboratory radiobiology is performed at room temperatures, which could have consequences for informing about the clinical situation. Since α- and β_o-inactivation processes exhibit qualitative as well as quantitative differences, the proportion of tumor cell killing by each mechanism at various fraction sizes is important, especially when altered fraction sizes are contemplated. The α/β_o ratio defines the dose at which there is equal killing by the different mechanisms. Since the fraction size used in the majority of radiotherapy protocols is much lower than the α/β_o ratio of the tumor clonogens, killing by the single-hit mechanism will dominate.

If exposure times exceed about 5 min for cells irradiated at 37°C, repair of sublesions of the β_0-mechanism will result in higher surviving fractions than predicted by the basic LQ equation. Consequently, dose-rate and overall treatment time will be important factors for predicting the effectiveness of hypofractionated protocols. For irradiations at 37°C, a parameter that accounts for the repair of sublesions during the radiation exposures should be included in the linear quadratic (LQ) equation to predict accurately for tumor cell killing. Reasonable estimates of these repair rates have been reported.

As well as the "quality" of the radiation used in cancer treatments, other physical factors that strongly impact tumor treatment response are the fraction size, the total dose and the overall treatment time of the prescriptions. It cannot be overstated that no experimental evidence suggests that the molecular lesion(s) responsible for α-inactivation of human tumor cells can be repaired by endogenous enzyme processes. Attempted repair of complex DNA lesions can produce the chromosome aberrations observed at the first cell division after radiation exposure (Chadwick and Leenhouts 1981). The lethal lesions associated with α-inactivation probably consist of several chemical changes to DNA in close proximity within individual and between adjacent chromosomal domains (see Chapter 6). On the other hand, the sublethal lesions responsible for β_0-inactivation can be completely repaired if given enough time as evidenced by the elimination of the quadratic component of cell survival curves when the radiation is delivered at low dose-rates. The DNA lesions associated with β_0-inactivation are relatively simple and have multiple pathways for enzymatic repair. Normal tissue complications can result from the inactivation of stem cells by both mechanisms but tissue "architecture" as well as damages to vasculature and other stromal components could also play important roles (see Chapter 9 and Stewart and Dörr 2012).

Radiobiology research has shown that when cells are irradiated at 37°C at very low dose-rates (0.02 Gy/min and less), most, if not all, cell killing will result from the α-inactivation mechanism (Bedford and Mitchell 1973; Hall 2000). At the dose-rates and fraction sizes of most current treatment protocols, sublethal damage repair of the β_0 mechanism is of little consequence for tumor response outcomes since cell killing is mainly by α-inactivation. The current practice of administering 1.8–3 Gy fractions for 5 days a week up to the total dose that is well tolerated by exposed normal tissues was derived empirically over many years of clinical practice. These treatment schemes obviously produce an adequate killing of tumor clonogens each day while allowing the cells in the necessarily exposed normal tissues to experience adequate repair of their β_0-damages. Since the growth fraction (proliferating/quiescent + proliferating) is higher in most tumors relative to normal tissues, α-inactivation is likely to be higher in the tumor cells. We currently teach that the α/β_0 ratios for tumor clonogens and acute tissue responses (possibly stem cell based) are ~10 and those for late tissue complications are ~3. This is an oversimplification and more representative α/β_0 ratios for different classes of tumors will be suggested in Chapter 7. The radiation dose to normal tissues outside the planned treatment volumes (PTVs) of current treatment plans should be significantly lower than those experienced in the studies that defined todays optimal treatments. Radiation oncology research should investigate alternative dose fractionation schemes to increase tumor treatment responses and to decrease normal tissue complications with prescriptions that are less stressful for patients—potentially, at a reduced cost.

The research of Elkind in the 1960s was most important for defining what is known as *sublethal damage repair* in irradiated mammalian cells. Figure 5.1 shows the surviving fractions of Chinese hamster cells exposed to a single dose of ~15.6 Gy compared to when that dose was administered in two almost equal fractions separated by various times (Elkind, Sutton-Gilbert, et al. 1965). In this experiment, cells were incubated at 24°C between the exposures so that the variations that arise from cell cycle progression would be minimized. It is evident that the fractionated treatments produced less killing and this was attributed to the enzymatic repair of sublesions in the cellular targets during the inter-radiation interval. This repair was shown to be complete within about 2 hours at 37°C. When cells are grown to a stationary phase, irradiated and plated immediately or at various times after treatment, the delayed plating results in reduced cell killing (Hahn and Little 1972). This effect was defined as *potentially lethal damage repair* and is simply a different experimental expression of the repair of sublesions in the DNA target that can produce lethal events. At the time of these experiments, the LQ formalism was under development and was not used to quantify these effects. It is now known that this process of radiation-induced sublesion repair is exclusive to the β_0-inactivation mechanism and can be mathematically described by the LQ formalism (see Chadwick and Leenhouts 1981).

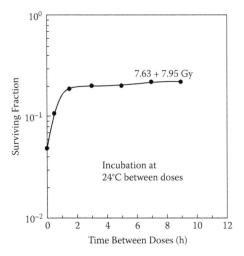

Figure 5.1. The surviving fraction of Chinese hamster cells irradiated acutely with ~15.6 Gy or when the dose is divided into two almost equal fractions separated by various times. The cells were held at ~24°C between the dose fractions to inhibit cell cycle progression. (Adapted and reproduced with permission from Elkind, M. M., Sutton-Gilbert, H., et al. 1965. *Radiation Research* 25:359–376.)

When clinical dose is fractionated over several days of treatment, the expected cell killing must account for the repair of β_0 sublethal damages between fractions. With each new fraction, cell killing will be expressed by the initial survival curve if sublethal damage repair is complete and tumor clonogens have redistributed within the cell cycle to their initial distribution. As described in Chapter 3, the difference in clonogen killing in different phases of their growth cycle by dose fractions of 2–3 Gy is quite small. The net tumor cell killing by fractionated protocols is then described by

$$SF = e^{-n\left(\alpha d + \beta_0 d^2\right)} \tag{5.1}$$

where

α and β_0 are the single-hit and double-hit inactivation parameters defined by the basic LQ equation

n = number of dose fractions

d = dose per fraction

Experiments reported in the early 1970s by Bedford and Mitchell (1973) showed the importance of sublethal damage repair when cells were exposed at lower dose-rates. The radiation sources used for many radiobiology studies in different laboratories are "retired" clinical x-ray units and produce dose-rates of 1–2 Gy/min at positions convenient for cell exposures. Cells irradiated at 37°C with dose-rates lower than 1 Gy/min exhibited reduced cell killing (see Figure 5.2). At dose-rates near 0.01 Gy/min, the resulting survival curves

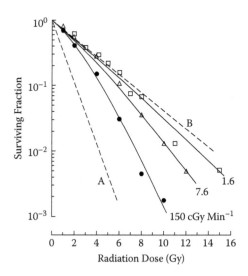

Figure 5.2. The surviving fractions of a human melanoma cell line irradiated at dose-rates of 1.5, 0.076 and 0.016 Gy/min. The data were fitted to the LPL model, from which the lines A and B are derived and which indicate conditions for no repair and full repair, respectively. (Reproduced with permission from Steel, G. G. et al. 1987. *Radiotherapy Oncology* 9:299–310.)

became exponential (α-inactivation only), indicating that the β_0-inactivation mechanism could be completely eliminated. At even lower dose-rates, surviving fractions were even larger, an effect attributed to cell proliferation during the extended exposures of several hours. The killing of clonogens in human tumors by brachytherapy protocols is assumed to be by α-inactivation only, but there will be large gradients in both dose-rate and killing effectiveness around the individual implanted radioactive seeds. To predict and understand the tumor response to altered fractionation schemes it is important to identify the extent to which each killing mechanism is in play.

The three human tumor cells lines of Figure 3.5 expressed significantly different intrinsic radiosensitivities accounted for by differences in α-inactivation during the interphase. When these cells were irradiated at 37°C at low dose-rate (0.015 Gy/min), the resultant survival curves became exponential on a semilogarithmic plot (see Figure 5.3) with $\bar{\alpha}$-inactivation parameters similar to those obtained from best fits to survival data generated at high dose-rate (1.4 Gy/min) (Chapman et al. 1999). For these three tumor cell lines, $\sqrt{\beta_0}$-inactivation rates were not significantly different. The experiments were performed with asynchronously growing cells, and whether or not the repair of sublethal damages associated with the $\bar{\beta}_0$ mechanism is constant at each position in the cell cycle was not thoroughly investigated. It is known that at least two important "checkpoints" occur in the cell cycle, one prior

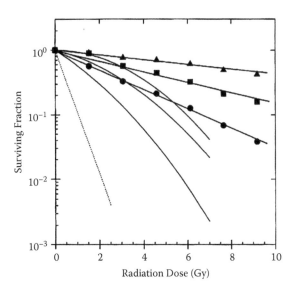

Figure 5.3. The surviving fractions of A2780 (circles), OVCAR10 (squares) and HT-29 (triangles) tumor cell lines irradiated at 37°C at a dose-rate of 0.014 Gy/min. The solid lines (without data points) indicate the survival of acutely irradiated asynchronous cells (from Figure 3.5, Chapter 3). The leftmost curve is for all cell lines at mitosis. (Reproduced with permission from Chapman, J. D. et al. 1999. *Radiation Research* 151:433–441.)

to DNA synthesis and the other prior to mitosis—the phases of the cell cycle where major conformational changes are required for DNA replication and chromosomal segregation, respectively (Kaufmann 1995; Dasika et al. 1999). Presumably, when lethal lesions and/or potentially lethal lesions are identified in DNA at these times, cell cycle progression pauses to allow for DNA repair enzymes to execute their functions prior to the DNA synthesis and compaction processes. Earlier studies had shown that cell proliferation was delayed by 1–2 hr after each Gray of radiation dose—times that allowed for DNA repair (Elkind et al. 1963). Whether or not repair can occur in the compacted DNA of chromosomes at mitosis has been an intriguing question but is difficult to answer since the elevated temperatures required for enzymes to function optimally also permit cells to progress rapidly from mitosis into the G_1-phase.

When human tumor prescriptions are delivered with single or multiple ports at dose-rates of 1–2 Gy/min, the treatment times to deliver a daily dose fraction are brief (usually 5 min or less). For most cells in human tumors, radiation killing is dominated by the α mechanism at clinical doses of 1.8–3 Gy; however, when prolonged exposures are anticipated and especially when hypofractionation is prescribed, the repair of $\overline{\beta}_o$-inactivation sublethal lesions during exposures should be taken into account.

Some early research with Chinese hamster cells investigated the repair of $\overline{\beta}_o$-inactivation sublethal lesions and DNA strand breaks as measured by alkaline and neutral sucrose gradient procedures (Dugle, Gillespie and Chapman 1976). Figure 5.4 shows survival curves for cells irradiated at a dose-rate of

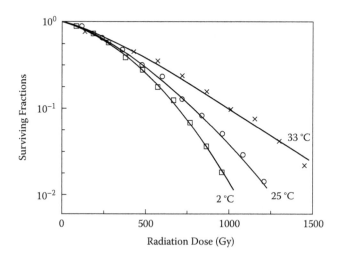

Figure 5.4. The surviving fractions of asynchronous Chinese hamster V79 cells irradiated at a dose-rate of 0.048 Gy/min at 2°C, 25°C and 33°C. The lines are best fits to Equation (5.2) from which first-order repair rates of sublesion damages were obtained. (Adapted and reproduced with permission from Chapman, J. D. and Gillespie, C. J. 1981. *Advances in Radiation Biology* 9:143–198, Academic Press.)

0.048 Gy/min at various temperatures (Chapman and Gillespie 1981). It is apparent that cell killing varied significantly and was reduced as the temperature increased. The basic LQ equation (Equation 2.2, Chapter 2) was then modified to include a first-order rate of repair of sublethal lesions (in this case, single-strand breaks) of the $\bar{\beta}_0$-inactivation mechanism (Gillespie et al. 1976; Chapman and Gillespie 1981) as

$$SF = \exp(-\bar{\alpha}D) \times \exp\left(-2\bar{\beta}_0 D^2 \left\{1 - [1 - \exp(-m)]/m\right\}/m\right) \qquad (5.2)$$

where
  $\bar{\alpha}$ and $\bar{\beta}_0$ are as previously defined
  D = radiation dose
  $m = kD/R$
  k = first-order rate of repair of β sublesions (temperature dependent)
  R = dose-rate

The data of Figure 5.4 and similar experiments were best-fitted to Equation (5.2) (conformed to the rate theory expression, $k(T) = f\, e^{-\delta H/R_0 T}$). The values for $\bar{\alpha}$- and $\bar{\beta}_0$-inactivation were from survival curves generated at low temperature or at high dose-rate and the solid lines in Figure 5.4 define the optimal repair parameters. Figure 5.5 shows an Arrhenius plot of these sublesion repair rates obtained from independent experiments performed on different days at various temperatures. These data yielded an enzyme activation coefficient (an important

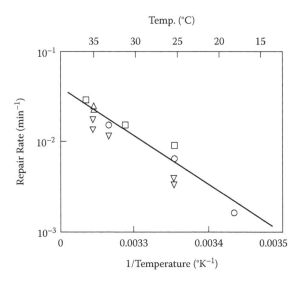

Figure 5.5. An Arrhenius plot of first-order repair rates of sublesion damage obtained from survival data fits to Equation (5.2) as a function of the reciprocal of absolute temperature ($°K^{-1}$). (Reproduced with permission from Chapman, J. D. and Gillespie, C. J. 1981. *Advances in Radiation Biology* 9:143–198, Academic Press.)

enzyme signature) of 24.3 ±1 kcal/mole, which agreed closely to the activation coefficient of 23.9 ±1.7 kcal/mole for the repair of single-strand breaks in these same cells measured by alkaline sucrose gradient procedures (Dugle and Gillespie 1975). Consequently, an LQ equation that incorporates a first-order coefficient for repair of the sublesions of the $\bar{\beta}_0$-inactivation mechanism (Equation 5.2) can be used to predict accurately for cell survival data generated at room and higher temperatures. This requires prior knowledge of the basic LQ parameters for the cells of interest. The first-order repair constants from the preceding study were 0.04/min at 37°C and 0.007/min at room temperature. The absolute values of these repair rates from cell survival assays at all temperatures investigated were about fourfold lower than the rates measured for single-strand break repair in the same cells as measured by sucrose gradient procedures. This suggests that single-strand break repair might constitute only one component, albeit a dominant one, of the sublethal damage repair measured by cell survival assays. Additional radiation-induced DNA strand break data generated at that time suggested that there could be a minor component of strand break repair that occurred at a lower rate (Dugle et al. 1976). More information about radiation-induced DNA lesions and their repair has accumulated over the intervening years using more sensitive assays (see Chapter 6) and enzyme complexes have been identified that can repair simple DNA double-strand breaks. How these enzyme systems relate to the sublethal lesions defined by the LQ equation is not completely understood. Nevertheless, the presence of unrepaired DNA DSBs or misrepaired cellular DNA at the time of DNA synthesis and mitosis could pose a lethal threat to clonogenic tumor cells.

Recently, others have recognized this same problem in that cell killing, after prolonged exposures, is less than that predicted by the basic LQ equation (Park et al. 2008; Kirkpatrick, Meyer and Marks 2008). This should come as no surprise and can be accounted for quantitatively but requires information about the intrinsic radiosensitivity of the cells of interest. Additional research should be performed to determine whether or not the first-order sublethal damage repair rates measured for Chinese hamster fibroblasts are similar to those for human tumor cells. In those cases where animal or human tumors are exposed to large dose fractions for prolonged times at 37°C, the basic LQ equation (Equation 2.2) will be found wanting. And when $\bar{\alpha}/\bar{\beta}_0$ ratios are extrapolated from clinical data of normal tissue and tumor responses to radiation, even less can be learned of the fundamental mechanisms involved.

DNA repair pathways that occur after cell irradiation will be described in greater detail in Chapter 6. It is now known that studies of sedimentation of DNA through alkaline and neutral sucrose gradients were of limited value since relatively large doses were required to produce DNA sizes that conformed to the laws of sedimentation velocity. Elkind coined the term *anomalous sedimentation* to describe the misbehavior of large DNA fragments within sucrose gradients (Elkind 1971). In the DNA strand break studies described before, radiation doses of 100–600 Gy were given to produce the single- and double-strand breaks that could be measured by these techniques (Dugle and Gillespie 1975; Dugle et al. 1976). At these doses, you can be sure that the $\bar{\beta}_0$-inactivation mechanism would be dominant. The extrapolation of these data to the dose range of radiotherapy was probably inappropriate. More recent techniques for investigating DNA strand breaks in tumor cells after doses of 1–20 Gy have been developed and some data generated with them will be presented in Chapter 6.

It should be noted that the radiation dose that produces equal amounts of cell killing by $\bar{\alpha}$ and $\bar{\beta}_o$ mechanisms is the much discussed $\bar{\alpha}/\bar{\beta}_o$ ratio where

$$e^{-\bar{\alpha}D} = e^{-\bar{\beta}_o D^2}$$

$$\text{or } \bar{\alpha}D = \bar{\beta}_o D^2 \tag{5.3}$$

$$\text{or } \frac{\bar{\alpha}}{\bar{\beta}_o} = D(\text{in Gy})$$

With daily doses of 1.8–3 Gy (the most common in current clinical use) and an $\bar{\alpha}/\bar{\beta}_o$ ~10 for tumor clonogens, the single-hit mechanism will dominate in producing the cell killing that results in tumor response. If hypoxic clonogens are the limiting cells, their $\bar{\alpha}/\bar{\beta}_o$ ratio might be even higher (see Chapter 6). When novel treatment prescriptions call for much larger fraction sizes (as in hypofractionation or radiosurgery procedures), the $\bar{\beta}_o$-inactivation mechanism will play a much larger or even the dominant role (Ling et al. 2010). To this day, most radiobiology studies have not given adequate recognition to the fact that these two different cell-killing processes (α- and β_o-inactivation) are operative. The next chapter will describe possible differences in the energy-depositing events and in the molecular targets associated with these cell-killing mechanisms. Dose fraction size becomes a critical factor in relating molecular mechanisms to cell killing involved in tumor response after radiotherapy, especially when hypofractionation is contemplated.

6 Ionizing Events, Molecular Targets and Lethal Lesions

In this chapter the physical concepts of "target theory" are reviewed in light of the different mechanisms of cell killing described by the linear-quadratic (LQ) equation. An improved understanding of the ionizing events in tumor cells is provided by microdosimetry and nanodosimetry. The majority of radiation dose deposited in the nuclei of tumor clonogens occurs as discrete events of low energy (60 eV on average) in mainly cellular water (70%–75% of a cell's mass), each of which will produce only a few free radicals that can diffuse 2–3-nm to DNA and produce potentially lethal lesions. These events are likely responsible for the simple molecular lesions associated with β_o-inactivation. A much smaller proportion of the delivered dose involves larger amounts of energy (~700 eV) deposited in much larger volumes of 10–30-nm dimensions where electrons are stopped. These can produce multiple and local DNA lesions. Electron track-ends, Auger cascades and charged-particle radiations deposit much of their energy in this manner, which can inflict several chemical changes in close proximity in the DNA target (intra- and interchromosomal) that are heterogeneous and irreparable. These radiation events probably account for α-inactivation. The DNA in mitotic cells and in condensed chromatin of interphase nuclei is most susceptible to these larger energy deposition events. As regards molecular targets, an improved understanding of cellular DNA in its various conformations throughout the cell cycle and in different cell compartments will be important for describing α- and β_o-inactivation mechanisms. Both mechanisms involve the indirect effects of water radicals in which OH˙ and molecular oxygen play decisive roles.

Since the radiation-induced killing of mammalian cells can be accurately described by two independent mechanisms governed by single-hit and double-hit dose kinetics, what is the evidence to suggest that they constitute qualitatively distinct processes? The data that inform about the "anatomy" of α- and β_0-inactivation mechanisms is scattered throughout the radiation chemistry and biophysics literature of the past 50 years. What made for interesting research was the identification of radiation chemical events that took place "in the wink of an eye" (10^{-12}–10^{-6} sec) that linked our understanding of energy depositions in cells with the tumor response and normal tissues complications that are expressed over days and months. Radiobiology research exploited techniques of microdosimetry, rapid-mix chemical analyses (pulse radiolysis) of DNA lesion induction, DNA lesion processing by repair enzymes, and cell kinetics of tumors and normal tissues along with clinical observations of tumor and normal tissue responses to understand and optimize radiotherapy treatment.

6.1 Time-Scale of Radiation-Induced Cellular Damages and Their Expression

The time-scale of radiation events was a popular topic during the 1970s and 1980s (Boag 1975; Chapman et al. 1975; Chapman and Reuvers 1977; Chapman and Gillespie 1981). Figure 6.1 shows the approximate time-scale (logarithmic)

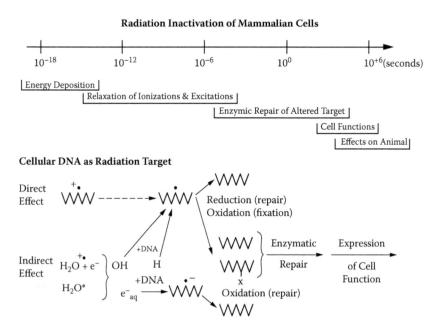

Figure 6.1. A schematic of the time scale of the physical, chemical, molecular and biological expressions of radiation damage in human tumor cells. (Reproduced with permission from Chapman, J. D. et al. 1974. *International Journal of Radiation Biology* 26:383–389.)

of radiation events in cell water measured by physical, chemical and biological assays and how they might be interrelated. Physical ionizations within tumor cells occur within the time it takes for fast moving electrons to pass nearby the electrical fields of orbital electrons in cellular molecules. This time is calculated to be 10^{-16}–10^{-14} sec. When the electrons ejected by that process diffuse from their creation site and interact with other cellular molecules—in particular, with more water—they produce a number of different free radicals (water radiolysis) at different frequencies and with different reactivity (Adams 1972). The positively charged molecule that is left behind will seek out neutralizing species. Of these reactive species, $H^{\cdot}$, $OH^{\cdot}$, e^{-aq} and H_2O_2 are considered to be the most potentially damaging. And since $OH^{\cdot}$ and e^{-aq} are produced at the highest frequency with G-values (number produced per 100 eV energy deposited) near 3.0, their reactions with several cellular molecules were quantified by pulse radiolysis techniques (Buxton et al. 1988). Biological molecules that reacted selectively with individual radical species were identified and their role in cell-killing mechanisms was investigated.

Chemical scavengers of $OH^{\cdot}$ were shown to protect against at least 65% and 30% of all the radiation killing of aerobic and hypoxic mammalian cells, respectively (Reuvers et al. 1973; Chapman et al. 1973; Greenstock et al. 1974). Studies with DNA in solution showed that $OH^{\cdot}$ could abstract hydrogen atoms from its sugar constituents (the most exposed portion of the molecule) and, when these sites reacted with molecular oxygen, DNA strand breaks resulted (Ward 1975). The kinetics of protection by $OH^{\cdot}$ scavengers of radiation-induced DNA strand breakage and cell killing were quite similar (Roots and Okada 1972, 1975; Chapman et al. 1974). N_2O was identified as an effective scavenger of e^{-aq} in radiation chemical studies. A product of these reactions was an additional $OH^{\cdot}$ in water. When N_2O was employed to scavenge e^{-aq} in irradiated mammalian cells, no radioprotection was observed. At room temperature, N_2O is a gas and its solubility in water was not high enough to produce the concentrations (near 1 M) required to scavenge (efficiently) other water radicals produced in irradiated cells. So whether or not the residual cell killing (<35%)—that not attributed to $OH^{\cdot}$—resulted from other diffusible water free radicals or direct ionizations in cellular DNA was not established by these studies. The reaction rates of $OH^{\cdot}$ and e^{-aq} with most cell molecules *in vitro* and *in vivo* are extremely rapid (in the order of 2–8 $M^{-1}s^{-1}$) and consequently a high molar concentration of any radical scavenger (~1 M) is required to produce effective radioprotection. The times of reactions of water radicals with target material in cells are 10^{-9}–10^{-8} sec. There are no known compounds that could be administered to patients at these very high concentrations (without toxicity) that could alter these initial and extremely rapid intracellular radiation processes. While dimethyl sulfoxide (DMSO) has been a useful tool for defining the role of indirect effects in irradiated cells *in vitro*, its use as a protector of normal tissues in radiotherapy is not feasible.

Another class of chemicals that showed radioprotective efficacy against radiation-induced killing was SH-compounds whose effects were observed with concentrations of only 1 mM (Vos and Kaalen 1962; Roots and Okada 1972). Cysteamine was shown to add to the intracellular pool of radical-reducing (hydrogen-donating) species and rapidly repair the lesions in DNA sugars produced mainly by the abstraction of hydrogen atoms by $OH^{\cdot}$ and possibly direct effect (Chapman et al. 1973).

This chemical "repair" process was shown to be in competition with molecular oxygen for the same free radical sites in cellular DNA (Figure 6.1). The reduction/oxygenation state within cells could be chemically modified so that chemical repair (reduction) or chemical damage (oxidation) could be maximized (Chapman et al. 1973). This research provided a concise chemical interpretation of the "oxygen fixation hypothesis" that had been described several years earlier (Alper and Howard-Flanders 1956). The times for these reactions within mammalian cells were 10^{-5}–10^{-3} sec and, consequently, only millimolar concentrations of SH-compounds and O_2 were required to modulate radiation cell killing.

A major research effort by radiobiologists and radiation oncologists in the 1970s and 1980s involved the identification of oxygen-mimicking (electron-affinic) drugs that could compete in this intracellular redox process to selectively radiosensitize resistant clonogens in tumors that lacked oxygen (in hypoxic microenvironments). It was anticipated that these drugs would not be rapidly metabolized and that their diffusion into hypoxic regions of solid tumors would be efficient. Metronidazole, misonidazole, etanidazole and nimorazole were tested in several clinical studies but showed little or no benefit as applied (Overgaard and Horsman 1996). In hindsight, this was probably due to the fact that the amount of drug that could be administered with each dose fraction (because of neurotoxicity) was much lower than those that produced the radiosensitizing effects observed with cells *in vitro*. As well, it is likely that not all the human tumors recruited to these clinical trials contained hypoxic cells that limited their treatment response. Figure 6.2 shows the radiation killing of

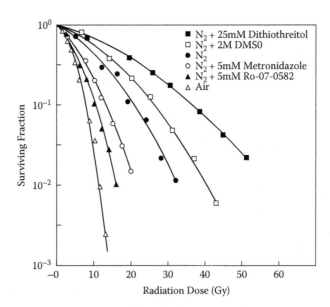

Figure 6.2. Survival curves for asynchronous populations of Chinese hamster V79 cells irradiated in air and hypoxia and in the presence of various chemical modifiers. The lines are best fits to the LQ equation.

Chinese hamster cells irradiated in various chemical environments that result in over a fivefold difference in intrinsic radiosensitivity.

Once stable chemical damages have been generated in cellular DNA by radiation, most can be restored back to normal by various enzymatic repair systems. Repair pathways for DNA single-breaks (SSBs) and double-strand breaks (DSBs) have been extensively investigated (Hoeijmakers 2001). They involve multiple steps by different enzymes, some operating as complexes, over times characteristic of enzyme function. To restore the majority of DNA damages in irradiated cells to normality, 10^3–10^4 sec of enzyme action are required—the time observed for the sublethal damage repair in cells (see Figure 5.1, Chapter 5). Cell viability in most radiobiology studies is determined by the "colony-forming" assay, which requires at least five cell divisions (10^5–10^6 sec) for expression. The time for most human tumor cell lines to progress around one cycle of proliferation is 20–30 hr. To determine if these repaired cells survive and lead to tumor regrowth in animal or human tumors takes several more rounds of proliferation (times up to 10^7 sec).

This brief description of the time-scale of radiation-induced reactions in tumor cells indicates that the relationship between the physical, the molecular, the biological and the tissue end-points is not explicit and assumptions are necessarily made. This current understanding was based upon various studies that employed techniques of theoretical and radiation physics, radiation chemistry, cellular radiobiology, molecular biology, enzymology, tumor cell biology and clinical radiation oncology (see book cover). It is expected that with additional techniques and research, new insights will improve our current knowledge of the time line of radiation-induced cell killing and tumor response to radiotherapy.

6.2   The Oxygen Effect and Oxygen Enhancement Ratio (OER)

For tumor response modeling it is important to employ reliable parameters for the intrinsic radiosensitivity of both aerobic and hypoxic cells, especially when the oxygenation status of the tumors under investigation is known. The potential role of hypoxia in tumor treatment failure was proposed by Gray et al. (1953) and the "oxygen fixation hypothesis" was elaborated by Alper and Howard-Flanders (1956). Additional research confirmed this discovery and measured the OER to be ~3.0 (Elkind, Swain, et al. 1965) for Chinese hamster cells. OER was defined as the ratio of doses administered to hypoxic versus oxygenated cells that produced equivalent biological effect (killing). This is a large modification of intrinsic cell radiosensitivity, and the potential consequences of hypoxic radio-resistance in cancer treatment have been the subject of much research. As will be shown in Chapter 10, a typical fractionated radiotherapy protocol for prostate cancer would produce about five logarithms more killing of aerobic cells compared to hypoxic cells. Such radioresistance could determine the difference between tumor cure and tumor regrowth if surviving hypoxic cells reoxygenate and proliferate. It should be noted that molecular oxygen is the most effective and the most reproducible sensitizer of radiation action in tumor cells that has been reported to date.

Of tumor modeling interest is whether or not the OERs for $\bar{\alpha}$ and $\bar{\beta}_o$ inactivation are the same. Early studies with asynchronous Chinese hamster cells irradiated as monolayers in glass petri dishes suggested that the OER for α-inactivation was significantly lower than that for β-inactivation (Chapman et al. 1975). Palcic and Skarsgard (1984) also concluded that the OER for Chinese hamster cells was lower at low doses of ionizing radiation. But "making" cells hypoxic for *in vitro* radiation studies is not a straightforward procedure. Our research employed the technique of flowing an oxygen-free gas through the air space of irradiation chambers for a period of time prior to exposures to dose. Cells were irradiated after equilibrium had been achieved between the gas phase and the attached cells that would be covered with some media. Some survival curves (see Figure 2.1, Chapter 1) were generated with a Perspex irradiation chamber that contained groups of shielded petri dishes, which could be sequentially positioned in the irradiation field after appropriate gassing times. Oxygen tension could be directly measured in the tissue culture medium adjacent to cells at the time of irradiation but not in various domains within the cells. After the irradiation chamber shown in Figure 2.2 (Chapter 2) was constructed, all subsequent studies that required hypoxic environments were performed with that system. Figure 6.3 shows the surviving fraction (SF) of mouse EMT-6 tumor cells irradiated in slowly stirred suspensions after a 10 Gy dose of ^{137}Cs γ-rays was administered during a continuous gassing procedure (1 L/min) with nitrogen + 5% CO_2 that contained known (measured)

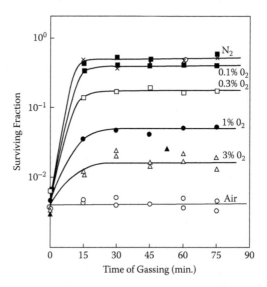

Figure 6.3. The surviving fraction of asynchronous mouse EMT-6 tumor cells irradiated with 10 Gy at various times after gassing at 1 L/min with nitrogen, 5% CO_2 and measured amount of oxygen. Cells were irradiated in chambers like that shown in Figure 2.2. (Reproduced with permission from Chapman, J. D. et al. 1991. *Radiation Research* 126:73–79.)

amounts of oxygen (Chapman et al. 1991). These data were published as part of a photodynamic therapy study where the essential role of oxygen in Photofrin II sensitization was demonstrated. It is apparent that 30 min of pregassing this chamber with the nitrogen gas mixtures produced plateaus of cell survival unique for each oxygen concentration. This dose (10 Gy) would produce almost equal killing by and $\bar{\alpha}$ and $\bar{\beta}_0$ mechanisms in aerobic cells. When G_1-phase and mitotic populations of human tumor HT-29 cells were irradiated with ^{137}Cs γ-rays in these same chambers at low temperature, the OER for mitotic cells (for which α-inactivation dominates) was 1.8–2.0 and significantly higher for G_1-phase cells (Stobbe, Park and Chapman 2002). The OER for asynchronous populations of mouse EMT-6 and human HT-29 tumor cells was ~2.8 at the 10% survival level. While the oxygen concentration dependency for G_1-phase cells after 45 min of pregassing with ultrapure nitrogen + 5% CO_2 followed the competition kinetics expected for an aqueous radiochemical process ($K_m = \sim 0.4\%$ O_2), the same response was not observed at these degassing times for mitotic cells. In previous photobiology studies that utilized this irradiation chamber, longer degassing times were required for oxygen in cellular lipids to equilibrate with those of the media and aqueous compartments of the cells (Chapman et al. 1991). If the studies with mitotic cells had utilized pregassing times of 2–3 hr and a larger OER has been observed, this would suggest that the radiation target (presumably DNA) in mitotic cells was in a lipid environment. Oxygen solubility is ~10 times higher in lipid relative to water and its diffusion coefficient is lower (Matheson and Rodgers 1982). This study of the OER of mitotic cells should be repeated with longer gassing times to determine if it is significantly different from that of interphase cells. A demonstration that lipid is a significant component of mitotic chromosomes would alter our understanding of where nuclear membrane lipid goes during mitosis.

Until further studies are performed, we recommend that OERs of 1.8–2.0 and 3.0–3.5 be employed for $\bar{\alpha}$- and $\sqrt{\beta_0}$-inactivation in tumor response modeling, respectively. These values will produce tumor control probability (TCP) responses that are slightly different from those that incorporate a constant value of OER = 3 for both inactivation mechanisms.

Koch developed a procedure for making cells hypoxic for radiobiological studies by growing them in glass dishes which were then placed in sealed stainless steel chambers for equilibration with different gas atmospheres. Multiple exchanges with gases of known oxygen concentration were used to establish equilibrium between in the cells and the gas phase (Koch, Kruuv and Frey 1973; Koch and Painter 1975). This procedure usually required a few hours to accomplish all the gas exchanges and the OER for radiation killing appeared to be constant over the survival curves, suggesting that the OERs for and $\bar{\alpha}$ and $\bar{\beta}_0$ mechanisms are the same. The details of these two procedures for producing hypoxic cells for radiobiology studies are presented to demonstrate that the research of tumor cell oxygenation status is not a simple matter. The different techniques probably produce different oxygen levels in the aqueous and lipid compartments of cells at different times. Early research showed how oxygen residing in plastic petri dishes influenced significantly the measurement of OER of attached cells (Chapman, Sturrock, et al. 1970). Much has been learned about acute and chronic hypoxia in animal tumors (Brown 1979; Chaplin, Durand and Olive 1986; Chaplin, Olive and Durand 1987) at the tissue pathology level, but oxygen solubility and its diffusion coefficient in intracellular domains of different chemical composition will make tumor response modeling even more difficult.

6.3 Radiation Events—The Role of Energy Density and Ionization Volume

The mechanisms by which photons and charged particles produce ionizations in cellular matter are well described by radiation physics (Johns and Cunningham 1983; Blakely et al. 1984; Mayles, Nahum and Rosenwald 2007). Accelerator-produced photons or protons will set in motion secondary electrons in tumor cells that will account for the majority of the cytotoxic effects. Clinical photon beams produce secondary electrons by the photoelectric effect, by Compton scattering and by pair production described by radiation physics (Johns and Cunningham 1983; Mayles et al. 2007). In fact, the majority of dose deposited in irradiated human tumors will come from Compton-scattered electrons, as dictated by the energy distribution of the primary photon beams and the low-Z composition of the biological tissues. Once secondary electrons have been released from cellular molecules (the most frequent being cellular water), they produce ionizations in other cellular molecules according to the Bethe–Bloch equation (Johns and Cunningham 1983; Mayles et al. 2007). The electric fields of electrons traveling at high speed will briefly interact with the electric fields of electrons in cell molecules through mainly glancing interactions producing excitations and ionizations. When ionizations occur, vacant electron sites (exhibiting a net positive charge) are created. These molecular ions then react with other cellular molecules, particularly water molecules, to gain back their lost electrons and return to their uncharged state via very rapid chemical processes.

The science of spatial ionization and excitation produced by various radiations at the molecular level is known as microdosimetry and nanodosimetry. While physical measurements of some rapid reactions can be made, most of our information about these processes comes from Monte Carlo simulations of charged particle stoppage. The last three decades of the last century saw many contributions to this field of study with frequent meetings of the involved scientists. Initially, the description of the physical events produced along charged particle tracks dominated this discourse. Over this same period of time, chemists were measuring the production of several products of water radiolysis, the lifetimes of these free radical species and their reaction rates with different cellular molecules (Adams 1972). The combination of these fields of research has now led to our current understanding of the intracellular pathways within the various cellular compartments (Figure 6.1).

Figure 6.4 shows the percentage of "first-collision" energy losses from 20 keV electrons passing through a thin layer of Formvar, a plastic whose density is the same as that of human tissue (Rauth and Simpson 1964). The most probable energy loss was 22 eV, while the average energy loss was about 60 eV. These are discrete

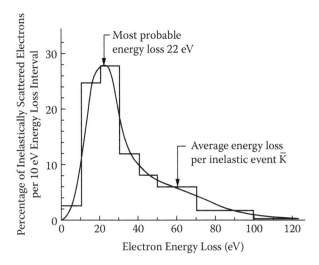

Figure 6.4. The distribution curve of the "first-collision" energy loss events for 20 keV electrons passing through a thin layer of Formvar, a tissue-equivalent polymer. The most probable energy loss is 22 eV and the average energy loss is 60 eV. (Reproduced with permission from Rauth, A. M. and Simpson, J. A. 1964. *Radiation Research* 22:643–661.)

events with the majority below 100 keV, a relatively small amount of energy. Radiation chemistry research produced G-values for the production of hydroxyl radicals (OH$^{\bullet}$), solvated electrons (e^{-aq}) and hydrogen peroxide (H$_2$O$_2$) in irradiated water of ~2.9, ~2.9 and ~0.65, respectively (Buxton et al. 1988). Again, G-value is the number of chemicals entities produced by 100 eV of deposited energy. For the majority of the events described here, at most 1–3 OH$^{\bullet}$, 1–3 e^{-eq} and 1 H$_2$O$_2$ could be produced in water at each ionization site. These free radicals would then diffuse to and react rapidly with surrounding water molecules or with other cell molecules in close proximity (within a few nanometers). The reaction volumes of these radical species are 4–5-nm diameter, which includes their diffusion distances. Radiation chemists refer to these low-energy events as "spurs" (Mozumder and Magee 1966a, 1966b). That research also identified short tracks (0.5–5.0 keV events) and spherical blobs (0.1–0.5 keV events) as other possible track entities that could have a higher specific energy than those of spurs. Spherical blobs were likely a different characterization of what we have identified as electron track-ends. This research prompted our modeling by Monte Carlo techniques the stoppage of 1 keV electrons in water, which showed average track lengths of 25-nm and average LETs of 40 keV/μm. If only the final 700 eV of energy is considered, the average track lengths are about 15-nm and their average LETs are 70 keV/μm (see Figure 6.5). Single particle events that span 10–30-nm dimensions could produce the irreparable damages associated with single-hit cell killing (Chapman 1980; Chapman and Gillespie 1981). The specific energy of these larger volume events is greater than that of spurs. This Monte Carlo code was written by C.J. Gillespie

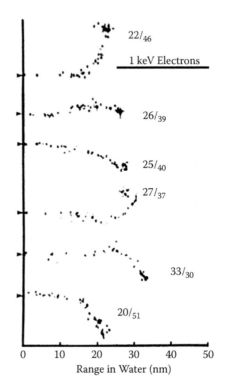

Figure 6.5. Two-dimensional Monte Carlo simulations of the stopping of 1 keV electrons in water. The large-type number on each track is its length in nanometers and the small-type number is its LET in kiloelectron volts per micrometer. (Reproduced with permission from Chapman, J. D. and Gillespie, C. J. 1981. *Advances in Radiation Biology* 9:143–198, Academic Press.)

for a Hewlett Packard Model 9845 computer that had the largest memory available at that time (64 kB) and required several minutes of time to generate a single plot. How times have changed!

The radiation killing of mammalian cells was known to result from both direct and indirect actions (Hall 2000). These terms refer to direct ionizations in the target molecule (DNA) and from secondary reactions of diffusible radicals with this target. The role of water radicals in the inactivation of mammalian cells was investigated with chemicals that could selectively react with specific free radical species (Chapman et al. 1973). The capacity of *tertiary*-butanol, ethylene glycol, *iso*-butanol and DMSO (chemicals that reacted selectively with $OH^\bullet$ at known rates) to radioprotect Chinese hamster cells was measured (Reuvers et al. 1973). Survival curves at that time were analyzed by the single-hit multi-target (SHMT) equation and alteration of the terminal slope (k) was used to define radioprotection. Both $\bar{\alpha}$- and $\bar{\beta}_o$-mechanisms of the LQ equation would be operative over the dose range used to obtain these k-values. A maximum radioprotection was observed only with DMSO since the other alcohols were toxic at

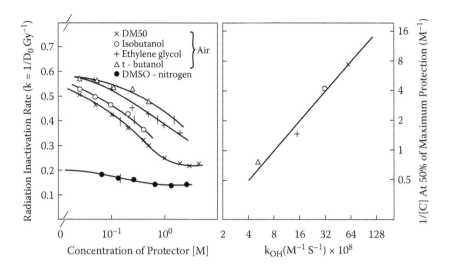

Figure 6.6. The radioprotective effectiveness of DMSO, *iso*-butanol, ethylene glycol and *t*-butanol on the radiation killing of asynchronous Chinese hamster V79 cells. Complete radioprotection was only achieved with DMSO (left panel). The reciprocal of the concentration of each radioprotector that produced one-half maximal protection is plotted against its reaction rate constant with OH$^\bullet$ in the right panel. (Reproduced with permission from Reuvers et al. 1973. *International Journal of Radiation Biology* 53:7125–7135.)

high concentrations (Figure 6.6, left panel). The vertical bars in Figure 6.6 (left panel) indicate the concentrations of these agents required to produce one-half the maximum radioprotection observed with DMSO. When the reciprocals of those concentrations (a measure of their protection effectiveness) were plotted against their absolute reaction rates with OH$^\bullet$ (from pulse radiolysis studies), the data conformed to a precise correlation (Figure 6.6, right panel). It was then reasonable to conclude that DMSO produced its radioprotection in mammalian cells by reacting with potentially damaging OH radicals prior to their reaction with radiation targets within cells. This study indicated that ~65% of all cell killing of these aerobic cells was via the indirect action of OH radicals. In another study it was found that only ~30% of cell killing in hypoxic environments resulted from OH radicals (Chapman et al. 1974).

The search for agents that could be used to scavenge e^{-aq} or H$_2$O$_2$ selectively in similar cell biology studies turned up nothing useful. The majority of the water radical scavengers used in radiation chemistry proved too toxic to mammalian cells at the concentrations that were expected to produce radioprotection. The enzyme superoxide dismutase, which reacts with H$_2$O$_2$ as an intracellular detoxifier, works at much slower enzyme speeds and could not compete in these initial radiation processes. Consequently, it is not known if the residual cell killing (less than 35% in aerobic cells) after radiation is by only direct effects or includes some effects of other diffusible radical species.

DMSO was then utilized to determine if OH radicals were involved in cell killing by both the $\bar{\alpha}$- and $\bar{\beta}_0$-mechanisms. Chinese hamster cells were synchronized by mitotic selection and some were incubated for 2 hr to produce populations of G_1-phase cells. Figure 6.7 shows survival curves for G_1-phase cells irradiated with 250 kVp x-rays (a) and spread-peak Ne-ions of maximal relative biological effectiveness (RBE) (b) under aerobic conditions at low temperature in the absence and presence of various concentrations of DMSO (Chapman et al. 1979). DMSO was found to protect equally against the radiation killing of these cells by both the low-LET and the high-LET radiations. When these and similar data produced with other charged-particle beams were best fit to the LQ equation and relative values of α and $\sqrt{\beta}$ plotted against DMSO concentration, the competition kinetics of this radiochemical action within cells was demonstrated (Figure 6.8). The quadratic parameters are often expressed as $\sqrt{\beta}$, which is proportional to the first power of dose, as is α. The data show that a significantly greater component of α-inactivation (80%–90%) results from the indirect action of $OH^\bullet$ than for $\sqrt{\beta_0}$-inactivation (~50%). In addition, the concentrations of DMSO required to protect against α killing were at least three times greater than those that produced protection against β killing. This indicates that the average diffusion distance of the lethal OH radicals to the molecular targets associated with single-hit killing are shorter (and/or their energy densities greater) than those associated with double-hit killing. Competition kinetic analysis indicated average $OH^\bullet$ diffusion distances of about 0.8 and 2.0-nm for

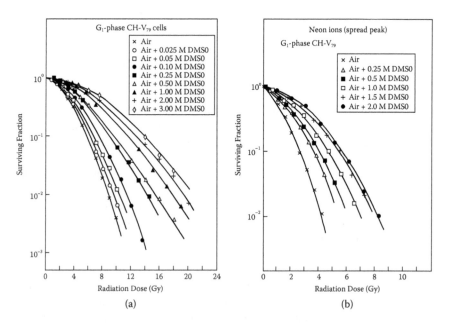

Figure 6.7. The radioprotection afforded G_1-phase populations of Chinese hamster cells by various concentrations of DMSO with 250 kVp x-rays (a) and spread peak neon ions. (b). (Adapted and reproduced with permission from Chapman, J. D. et al. 1979. *Radiation and Environmental Biophysics* 16:29–41.)

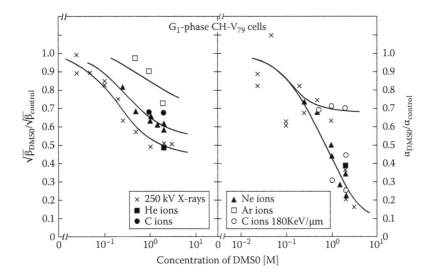

Figure 6.8. The effectiveness of various concentrations of DMSO (ratios of inactivation parameters) on cell killing by the $\sqrt{\beta}$ (left panel) and α (right panel). The studies were performed with G_1-phase Chinese hamster V79 cells. (Adapted and reproduced with permission from Chapman, J. D. and Gillespie, C. J. 1981. *Advances in Radiation Biology* 9:143–198, Academic Press.)

α- and β-inactivation, respectively (Roots and Okada 1972; Chapman et al. 1976). These studies highlight the importance of radiation chemical processes in the radiation killing of tumor cells.

The distributions of ionizations and excitations produced by charged particles as they traverse an aqueous environment can be visualized in two-dimensional plots generated by Monte Carlo simulations of the fundamental processes described by the Bethe–Bloch equations. A major thrust of this research was to model the interactions of protons and heavy-charged particles in living cells (Krämer and Kraft 1994; Emfietzoglou et al. 2000; Plante, Ponomarev and Cucinotta 2011; Dingerfelder 2012). The density of energy deposition along the tracks is defined as LET, usually in units of kiloelectron volts per micrometer and increases as the particles slow down and eventually are stopped (the Bragg peak). Similar two-dimensional Monte Carlo simulations were produced to "visualize" the excitations and ionizations in tracks of 1 keV electrons as they are stopped in water or within cells (see Figure 6.5). As electron energy was degraded below 700 eV, the probability of side- and backscatter increased significantly and the ionizations (spurs) became more closely clustered. This resulted in larger volumes of higher energy deposition at the electron track-ends relative to the spurs produced by the energetic electrons. These larger volume events could potentially produce multiple damages in close proximity (nanometer dimensions) within cell chromatin. They were described as electron track-end lesions for which the acronym ETELs would apply (Chapman 1980). The ionizing energy of a spur (on average about 60 eV) could produce the single sublesions in cellular DNA that are largely reparable, and that of an electron track-end (700 eV) could produce about 10–15 heterogeneous molecular damages in close proximity within single and/or adjacent chromosomal domains that could be irreparable.

Research into the radiochemical production of lesions in cellular DNA prompted John Ward to define similar events as multiple lethally damaged sites (MLDSs; Ward 1982). These are not single chemical entities and, when produced by electron track-ends, are relatively rare events that are difficult to identify among the much more frequent SSBs and DSBs produced by spurs. These larger volume, complex lesions are likely responsible for many of the intra-and interchromosomal aberrations observed at the first cell division after radiation exposure that correlate with measured cell lethality (see Chadwick and Leenhouts 1981). Goodhead defined similar molecular events produced by various charged particles as "clustered lesions" (CLs) (Goodhead 1994). These large-volume events and damages in cells were predicted by microdosimetry, radiation chemistry studies of cellular DNA and cell survival curve analyses. It should be noted that the deposition of energy by 100 to 1000 eV stopping electrons is not understood at a fundamental physics level as well as that of electrons of higher or lower energy. Electron events at these energies compose a relatively small portion of total absorbed dose and their experimental investigation may require novel approaches. The recent research by Sanche and colleagues (Michaud and Sanche 2003; Sanche 1995, 2008) has added important understanding of the physics and chemistry of low-energy electrons, particularly those of 100 eV and less. These data have revealed novel processes by difficult laboratory procedures that should be extended to electrons of higher energies. The Monte Carlo simulation of energy deposition from various charged particles can now be visualized by computer programs with colored, three-dimensional graphics (Plante et al. 2011). But these representations will only be as good as the underlying physical principles that are input into their Monte Carlo programs.

Tumor cell killing can result from at least two different physical events (distinguished by different volumes) that produce damages in cellular DNA that are qualitatively distinct. The vast majority of radiation dose is dissipated in cells as low specific-energy events (spurs) with interactive volumes of about 4–5-nm diameter and result in simple DNA lesions. A much smaller proportion of total dose is dissipated in electron track-ends with interactive volumes of 10–30-nm and possibly with higher specific energy that produce ETELs, MLDSs or CLs. A therapy dose fraction of 2 Gy will produce in tumor cell nuclei (~6 μm diameter) some 23,500 spur events and about 200–300 electron track-ends. (This partitioning of total dose between only two events is recognized as an oversimplification.) Since DNA comprises less than 15% of the mass of a tumor cell nucleus, only a portion of these ionizing events would be expected to produce DNA damages by direct effect. It is the indirect effect of mainly OH$^{\bullet}$ produced by electron track-ends and spurs that will produce the vast majority of molecular damages associated with α- and β_{o}-inactivation, respectively.

Fluorescent markers now make it possible to distinguish individual chromosomes at mitosis and chromosome domains within interphase nuclei (Pearson 1972;

Muller and Wienberg 2006). How these chromosome domains are embedded in the nuclear membrane for order and stability and what conformational changes they undergo at replication and division is an ongoing study of molecular biology and biophysics. There is much more to be learned about the tertiary structure of chromatin throughout the cell cycle that could have important consequences for understanding both the α- and β_o-inactivation mechanisms and how they relate to cancer therapy.

6.4   The Molecular Target(s) for Cell Inactivation

That cellular DNA is the target for the radiation killing of tumor cells has been deduced from several lines of research. These include exploiting radiations of different ionization density and/or techniques that produced selective deposition of radiation energy within specific cellular macromolecules and compartments.

The early studies (introduced in Chapter 4) with cyclotron-produced α-particles and deuterons by Barendsen et al. (1963, 1964, 1968) and heavy-charged particles by Todd et al. (1967, 1974) yielded inactivation cross sections for mammalian cells of ~40 and 100 μm^2, respectively. These cross sections agree closely with the nuclear cross sections of the same cells as measured by microscopic procedures. It was concluded that the passage of a single high-LET particle through the cell nucleus was sufficient to produce cell death as measured by colony formation. Cole reported on cell killing by electron beams of 50 and 10 keV (Cole et al. 1974; Cole et al. 1975). The lower energy electron beam penetrated only 1/10 to 1/3 into the cell nuclei and the resultant cell killing was 3–10 times more efficient than that observed with fully penetrating (50 keV) electron beams. These studies also showed that the low-energy electrons were more efficient in producing single-strand and double-strand breaks in cellular DNA. Munroe (1970) irradiated selected portions of individual mammalian cells with α-particles from a polonium-tipped microneedle. He showed that up to 250 Gy to only the cell cytoplasm had little or no effect on cell proliferation. On the other hand, irradiation of the cell nucleus with 10 Gy resulted in cell inactivation through the formation of giant and multinucleated cells and, ultimately, by cell lyses. Munroe concluded that the targets for radiation killing of mammalian cells were in the nucleus and possibly at the nucleus periphery. The most efficient cell killing was observed when electrons and α-particles penetrated only 1 to 2 μm into the nuclei, indicating that the periphery of the nuclei might contain ultrasensitive radiation targets.

Marin and Bender (1963a, 1963b) measured cell inactivation produced by decays of [^{3}H]-thymidine and [^{3}H]-uridine that had been metabolically incorporated into the DNA and RNA of cells, respectively. The majority of the dose derived from the tritium decays in DNA was to the cell nucleus while the majority of the dose from the tritium decays in RNA was to the cell cytoplasm. The decays in the cell nucleus were much more effective in cell killing than decays in the cell cytoplasm. Burki et al. (1973) compared the effectiveness of the decays of ^{125}I and ^3H incorporated into the cellular DNA of Chinese hamster and mouse leukemia cells. For both cell lines, disintegrations of ^{125}I were much more effective per unit dose than those from ^{3}H. ^{125}I and ^{123}I decays produce six electrons (on average) with discrete energies of 0.5 to 34 keV, of which three have the lowest energy (Burki et al. 1973; Nikjoo, Emfietzoglou and Charlton 2008). This process is called an Auger cascade and the damage(s) produced by such events could be similar to those produced by electron track-ends (Yasui, Hughes and DeSombre 2008). As described in Chapter 4, the decays of ^{125}I and ^{123}I incorporated into cellular DNA produce cell killing mainly by the indirect effect of OH$^•$ generated in the surrounding water (Walicka 1998a).

Recent studies with various radiation microbeam systems have confirmed that the cell nucleus contains the molecular targets that are most sensitive to the killing effects of ionizing radiations (Prise 1998; Seymour and Mothersill 2004). These techniques have also facilitated the study of novel "bystander effects" of radiation that are briefly discussed in Chapter 8.

So if cellular DNA is the major molecular target for tumor cell killing, what is its structure during the cell growth cycle and does its conformation influence a cell's response to radiation? Tetraploid mammalian cells contain about 2.3 m of double-stranded DNA that is "packaged" into two complements of 23 chromosomes of variable sizes. This linear genetic code is "folded" by sophisticated molecular processes to fit within the cell nucleus of ~6-μm diameter. Figure 6.9 shows a schematic of the several stages of chromatin packing (condensation) that can be observed in mammalian cells by conventional and electron microscopic procedures. Between 180 and 190 base pairs of the DNA double helix (2-nm diameter) are "wrapped" around a complex of proteins called histones (H2A, H2B, H3 and H4) to produce nucleosomes of ~10-nm diameter that appear as the "beads-on-a-string" structure (Kornberg 1974; Kornberg and Klug 1981). These nucleosome strings are coiled further into a larger (possibly helical) structure of 30-nm diameter, forming the most common chromatin structure observed in interphase cells by electron microscopy (see Figure 6.10). These 30-nm fibers bind to and are anchored in lipid/protein sites in the nuclear membrane and extend out into the nuclear milieu to form what are observed as chromatin loops. These loops of DNA associate with proteins in each chromosome domain to create centromeres that, along with the nuclear skeletal protein spindles, facilitate the separation of chromatin to opposite poles at mitosis. During mitosis, the histones of these chromatin loops are chemically altered to promote their further folding into the chromosome structures observed by light microscopy. All chromatin complexes will be bathed in a nuclear milieu consisting of soluble proteins and water. It is mainly the sugar components of these DNA structures that are most accessible to this milieu and will selectively sustain radiation damages by diffusible radicals of water radiolysis.

Interest in chromatin structure as a factor in radiosensitivity has attracted biophysical research for several years since it was shown that mitosis, where DNA is maximally compacted into chromosomes, is the most radiosensitive position in the cell cycle (Terasima and Tolmach 1961; Sinclair and Morton 1963). More recent studies indicate that mitotic cells express radiation killing primarily by the α-inactivation mechanism and with a sensitivity close to that of DNA repair-deficient cells (Chapman et al. 1999). The majority of radiation chemistry research that identified radiation-induced DNA lesions utilized reagent "naked" DNAs

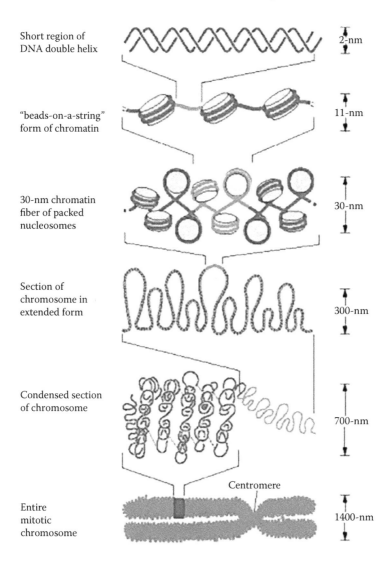

Short region of
DNA double helix
2-nm

"beads-on-a-string"
form of chromatin
11-nm

30-nm chromatin
fiber of packed
nucleosomes
30-nm

Section of
chromosome in
extended form
300-nm

Condensed section
of chromosome
700-nm

Centromere

Entire
mitotic
chromosome
1400-nm

Figure 6.9. The mechanisms by which the ~2-m length of DNA in a diploid human tumor cell is compacted to facilitate its residence in a cell nucleus of about 6-mm diameter. (Reproduced with permission from Alberts et al. 2002. Garland Science/Taylor & Francis LLC.)

irradiated in aqueous solution. Under these conditions, the radicals produced in the solvent could readily diffuse to the exposed parts of the 2-nm diameter structure. The condensation of cellular DNA with histones and lipids will provide significant protection against such indirect effects (Takata et al. 2013). With the exception of those portions undergoing active transcription or replication, DNA will be coiled in 30-nm diameter nucleosome fibers throughout the cell cycle.

This suggests that when further chromatin compaction occurs at the time of mitosis, some radiosensitive DNA/protein or DNA/lipid complex may be the target for α-inactivation. Whether or not the mechanisms for α-inactivation observed for cells in interphase and at mitosis are the same remains to be determined. Some studies have suggested an association between the percentage of chromatin in compacted form in interphase cells and their radiation killing by α-inactivation (Chapman et al. 1999; Chapman, Stobbe and Matsumoto 2001). A large number of studies of DNA structure and intrinsic cell radiosensitivity were recently reviewed by Lange, Cole and Ostashevesky (1993). That research did not distinguish between α- and β₀-inactivation mechanisms but did provide a wealth of knowledge about DNA association with the protein cytoskeleton of the nucleus and with nuclear membrane. Figure 6.10 shows a schematic of the solenoid model of coiled nucleosomes and the expected ionizations from an energetic electron (upper track) of 3 keV/μm of LET and those from an electron track-end (lower track) of over 30 keV/μm. It is apparent that the electron track-end could produce chemical damages in the DNA of multiple nucleosomes that reside within single or adjacent chromosome domains.

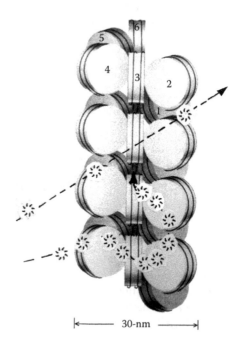

|←——— 30-nm ———→|

Figure 6.10. A schematic diagram of the solenoid model of nucleosome coiling into 30-nm diameter strands that are the most prevalent chromatin structure observed in electron micrographs. Superimposed on that structure are the ionizations expected from a fast moving electron (upper track; on average only two 60 eV events per 20-nm) and from an electron track-end (lower track; on average over twelve 60 eV events per 20-nm). The star symbols represent discreet 60 eV events, the average energy of the first energy events described in Figure 6.4.

Condensation (compaction) of chromatin in cells can be visualized and quantified by both light- and electron-microscopic techniques (Biade et al. 2001; Chapman, Stobbe and Matsumoto 2001). Gray-tone density distributions of 1% osmium tetroxide staining (nucleoprotein) were measured in thin section pixels of interphase nuclei (excluding the cell nucleolus) of interphase tumor cells and compared to density distributions of chromosome and non-chromosome regions (see Figure 6.11) of mitotic HT-29 cells (Chapman, Stobbe and Matsumoto 2001). The cell fixation, section thickness and staining procedures were identical for

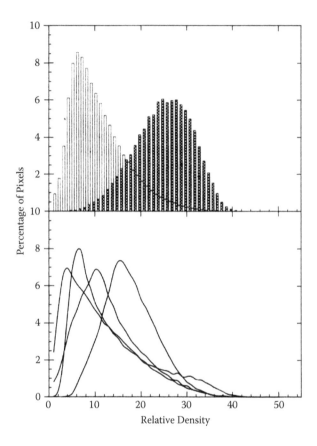

Figure 6.11. The distribution of relative densities of 1% osmium tetroxide staining (nucleoprotein) in 40 × 40-nm pixels of 80-nm sections of human tumor cells. The upper panel shows the densities of pixels in chromosomal (right) and non-chromosomal (left) regions of mitotic HT-29 cells. Densities of 17 units and greater were deemed to be compacted chromatin. The lower panel shows distributions of similar densities from the interphase nuclei (outside the nucleolus) of HT-29, OVCAR10, PC3 and A2780 human tumor cells (from left to right distributions). Values above 17 units were deemed to be compacted chromatin. (Reproduced with permission from Chapman, J. D., Stobbe, C. C., et al. 2001. *American Journal of Clinical Oncology* 24:509–515.)

all samples. When the ratio of pixel densities (about 40 × 40-nm and 80-nm thick) with "chromosome-like" density relative to non-chromosome density was plotted versus the α-inactivation parameter of these cells, a strong correlation was observed (Figure 6.12). These studies should be expanded to include a broader range of tumor cell lines to determine if the procedure could become a predictive assay of intrinsic radiosensitivity in human tumor biopsy specimens. Similar measurements of interphase pixel density distributions in repair-deficient cells (CHO-xrs5) showed portions of their chromatin in compacted form adjacent to their nuclear membranes (Chapman et al. 1999). This was not observed in wild-type (CHO-K_1) cells. This study suggested that the defective DNA repair functions in the radiosensitive cells could have as much to do with the chromatin structure at the time of radiation exposure as with postradiation DNA repair. Mitotic cells were found to have many more SSBs and DSBs in their DNA relative to G_1-phase cells, which suggests that DNA breakage and repair are normal functions in dividing cells (Stobbe et al. 2002). The processing of chromatin as it undergoes its conformational changes around the cell cycle is an active field of current research that could provide molecular leads for clinical predictive assays and for modulated radiotherapy prescriptions. And novel fluorescent stains for histone chemical alterations might contribute important information about these radiation targets in cells.

Chromosome formation at mitosis involves the enzymatic alteration of nuclear histones by phosphorylation, deacetylation and methylation (Gurley et al. 1978; Nurse 1990; Pollard et al. 1999; Costello et al. 1996). Chemicals that produce

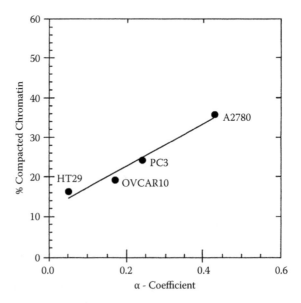

Figure 6.12. The percentage of osmium tetroxide stained pixels of 17 units and higher from Figure 6.10 (lower panel) versus the α-inactivation coefficient of asynchronous cells. (Reproduced with permission from Chapman, J. D., Stobbe, C. C., et al. 2001. *American Journal of Clinical Oncology* 24:509–515.)

premature condensation of chromatin in interphase mammalian cells have been exploited for studies of radiation-induced chromosome aberrations (Yamashita et al. 1990; Roberge et al. 1994). These chemicals were consequently investigated to determine if they (1) could be exposed transiently to human tumor cells *in vitro* without toxicity, (2) could produce radiosensitization by enhancement of α-inactivation and (3) if any measured radiosensitization correlated with specific histone chemical alterations and/or chromatin compaction (Biade et al. 2001; Price et al. 2004). Okadaic acid (a protein phosphatase inhibitor), fostriecin (a topoisomerase inhibitor) and trichostatin A (a histone deacetylase inhibitor) all produced significant sensitization of oxygenated cells at concentrations that were not toxic after short exposures *in vitro* (Biade et al. 2001). These drugs also increased chromatin compaction that could be visualized by electron microscopy of the nuclei of treated cells and by increased expression of phosphorylated histone 3 (measured with fluorescent antibodies). Phosphorylated histone 3 was known to be a marker for compacted chromatin in mitotic chromosomes (Nurse 1990). The data were consistent with the hypothesis that condensed chromatin in interphase tumor cells correlates with increased cell killing by mainly α-inactivation. Price synthesized several novel analogues of cantharidin (an inhibitor of protein phosphatase 1 and 2A) and tested them for radiosensitizing activity, chromatin compacting efficiency and phosphorylated histone 3 expression in DU-145 prostate cancer cells (Price et al. 2004). The optimal drug was designated LS-5 and exhibited excellent radiosensitizing activity (Figure 6.13) that correlated

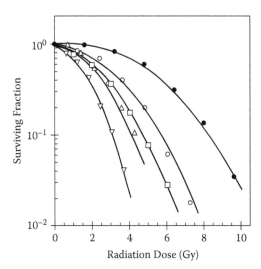

Figure 6.13. The surviving fractions of G_1-phase HT-29 human tumor cells after a 2-hr exposure to various concentrations of LS-5 at room temperature. These drug treatments produced no toxicity but were seen to potentiate cell killing. Control (solid circles), 0.2 mM (open circle), 0.5 mM (open square), 1.0 mM (open triangle) and 2.0 mM (inverted open triangle). (Reproduced with permission from Price, W. A. et al. 2004. *International Journal of Radiation Biology* 80:269–279.)

with chromatin compaction and histone 3 phosphorylation. Such drugs could potentially have a role in clinical radiotherapy if their sensitizing effect could be made selective to tumor versus normal cells. At this time, animal tumor studies have not yet been performed with LS-5.

While the phosphorylation of histone 3 was used in the preceding study as a marker for chromosomal-like DNA, other modified histones also can play a major role. Histone 2A becomes acetylated when chromatin unwinds its nucleosome structure for transcription and replication (Sharma, Singh and Almasan 2012). It then becomes deacetylated prior to returning to its normal nucleosome form. The presence of radiation-induced DSBs rapidly results in the phosphorylation of histone H2A variant H2AX (Sharma et al. 2012). Its expression has become a popular assay for identifying radiation-induced damage in cellular DNA and for monitoring its repair (Paull et al. 2000). Why the volumes of the irradiated cells that express histone H2AX as detected by fluorescent microscopy are larger than the volumes of chromatin damage expected for α or β physical events (up to 20-nm) is not clear.

There is strong evidence to suggest that the tertiary structure of chromatin in the nucleus of human tumor cells plays an important role in radiation killing by the α-inactivation mechanism. Cells appear to be most radiosensitive when the solenoid coils of nucleosome are in close proximity. Whether or not regions of condensed DNA within cells constitute larger cross sections for electron track-end damage remains to be confirmed. The possibility that some DNA/lipid structure at the nuclear membrane is the molecular target for α-inactivation is strongly suggested. And the role of DNA repair enzymes in processing cellular chromatin throughout the cell cycle and, in particular, at mitosis is not clearly understood. These are areas of research that are uniquely suited for those with skills in microdosimetry, molecular biology and chromatin biophysics. And an understanding of the molecular targets associated with cell responses at low dose are of concern for those that are addressing the potential risks of human space travel (Belli, Sapora and Taboccini 2002).

6.5   Lesions Produced in Cellular DNA by Radiation

Throughout the final 30 years of the twentieth century, radiation chemists and biophysicists diligently measured the frequency of various chemical alterations produced in DNA by exposures to ionizing radiations (Ward 1988; Olive 2000). In some chemistry studies, double-stranded DNA that had been purified from various biological sources was irradiated in aqueous solution and characterized. DNA SSBs, DSBs, specific base damages, interstrand and intrastrand DNA cross-links, DNA–protein cross-links and other types of damage were detected (Ward 1988; Moiseenko et al. 1998). When DNA was irradiated in mammalian cells and then characterized after its extraction, the amounts of specific damages measured were much lower than those observed with "naked" DNA, and their relative frequencies were different (Olive 2000). Table 6.1 shows the approximate number of specific DNA damages produced within tumor cell

Table 6.1. Approximate Number of Specific DNA Damages Produced in Tumor Cell Nuclei[a] by a Radiation Dose of 2 Gy

Radiation event	Frequency/2 Gy
Ionization sites	~23,500
Electron track-ends	<500
DNA base damages	>3000
DNA SSB	~2000
DNA DSB	~80
DNA–DNA and DNA–protein cross-links	~100

[a] ~6-μm diameter.

nuclei after exposure to 2 Gy of low-LET radiation. Since this radiation dose kills only 20%–80% of most human tumor cells (see Chapter 7), it is obvious that all these lesions cannot be lethal. In fact, the vast majority of SSBs and DSBs can be repaired by enzyme complexes that have been identified (Radford 1985; Hoeijmakers 2001).

DNA strand breakage research has progressed from the techniques of sucrose gradient sedimentation (see Chapter 5) to pulsed-gel electrophoresis assays and on to those that utilize single cells, such as the single-cell gel electrophoresis assay (the Comet assay). The Comet assay was first described by Ostling and Johanson (1984) and further refined and standardized for measuring DNA SSBs and DSBs with alkaline and neutral gel electrophoresis techniques, respectively, by Olive and colleagues (Olive et al. 1990; Olive and Banath 1993). The technique involves embedding single cells (control and irradiated) in agarose media on microscope slides prior to lysing with an alkaline or neutral buffer and performing one-directional electrophoresis at a low voltage for specific times (Stobbe et al. 2002; Price et al. 2004). The slides are then stained with a DNA-specific fluorescent dye, images of individual cells acquired and the amount of DNA and the distance that it migrates from the position of the embedded cell quantified. These data are then used to compute "tail moments," which are defined as the integral of amount of DNA (fluorescence intensity) × extruded distance. The tail moment units reported by different research groups will not be identical but intraexperiment variations can be instructive. Additional studies are required by each laboratory to correlate their units of tail moment to absolute numbers of SSBs and DSBs. This Comet assay revealed (Figure 6.14) that the DNA in unirradiated mitotic cells contained significantly more SSBs and DSBs than cells in the G_1-phase (Stobbe et al. 2002). This indicates that the process of chromatin compaction into chromosomes at the time of cell division requires DNA strand breakage (possibly at specific sites) to facilitate its separation and segregation to daughter cells. How these "natural" DNA breaks interact with radiation-induced breaks to express increased radiation sensitivity at mitosis is not yet understood. Endogenous DNA repair enzymes in mammalian cells are likely to accomplish these DNA breakage and repair processes.

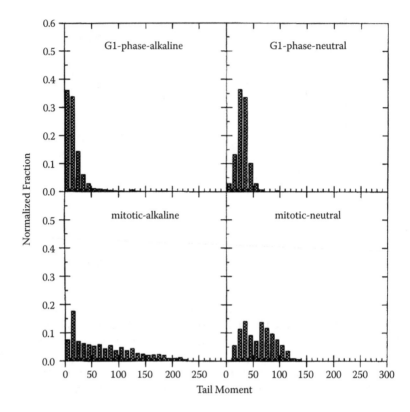

Figure 6.14. The distributions of tail moments (integral of amount × distance migrated) of DNA in unirradiated G_1-phase (upper panels) and mitotic HT-29 (lower panels) tumor cells. The left panels are for electrophoresis in alkaline buffer (SSB) and the right panels are for electrophoresis in neutral buffer (DSB). (Reproduced with permission from Stobbe, C. C. et al. 2002. *International Journal of Radiation Biology* 78:1149–1157.)

Mutant mammalian cells with specific defects in repair enzyme function could contain chromatin in interphase cells that already contains breaks that contribute to their hypersensitivity to radiation inactivation by the α-mechanism. Comet assays also showed (Figure 6.15) that the DNA in mitotic cells was more sensitive to the radiation-induced single-strand breakage than were G_1-phase cells (Stobbe et al. 2002). The conformation of cellular DNA in mitotic and interphase tumor cells appears to play an important role in their expression of DNA damages and intrinsic radiosensitivity.

The compaction of the ~30-nm diameter coils of nucleosome fibers into chromosomes at mitosis is accompanied by the hyperphosphorylation of histone 1 and the phosphorylation of histone 3 (Gurley et al. 1978; Wolffe 1998). Histone modulation throughout the cell cycle is a complex process that involves protein (serine threonine) phosphatases (PPs), kinases, acetylases,

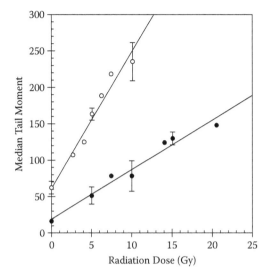

Figure 6.15. The tail moments for alkaline comets of mitotic and G_1-phase HT-29 cells exposed to various doses of ^{137}Cs radiation. Mean values ± SE are plotted for those cases where three or more independent determinations were obtained. (Reproduced with permission from Stobbe, C. C. et al. 2002. *International Journal of Radiation Biology* 78:1149–1157.)

deacetylases, methylases and demethylases that can regulate gene expression during interphase and effect chromosome formation for mitotic division (Roberge et al. 1994).

It is interesting to note that β-inactivation (for which sublethal damage repair can be complete) varies only slightly for the different radiations investigated and for most human tumor cell lines (see Chapters 7). The "simple" DNA damages associated with this cell killing probably result from those most frequent energy depositions described as "spurs." These events will produce some direct ionizations in cellular DNA in proportion to the DNA density in the nucleus (about 15%) but most will occur in the aqueous portion (about 70%) of the cell. It is these simple damages that can readily undergo repair and will account for the effects of dose fraction size and dose-rate in tumors and the interfraction recovery of normal tissues in treatment protocols. The demonstration that the kinetics of DNA SSB repair and sublethal damage repair in Chinese hamster cells are similar provides evidence for this assumption (see Chapter 5).

That lethal lesions in human tumor cells result from radiation events of different volume (those of "spurs" versus electron track-ends) provides a basis for the diverse expression of radiation damage observed as chromosome aberrations. Specific chromosome aberrations have been related to single-hit and double-hit radiation events by Chadwick and Leenhouts (1981). The introduction of multiple chemical lesions in local regions of cellular chromatin followed by attempts at their repair could explain the chromosome and chromatid aberrations that are visualized at the subsequent mitosis. The deposition of track-ends in chromatin at the boundary of two adjacent chromosomal domains in interphase nuclei can explain chromosomal interchanges. But since no molecular assay for ETELs, MLDSs or CLs has been devised, it is not yet possible to validate this proposed mechanism for the single-hit mechanism of the LQ equation experimentally. Attempts to simulate the radiation chemistry of energy depositions and their inter-actions with cellular DNA to produce the measured lesions have been an ongo-ing effort by several research groups (Chatterjee and Holley 1993; Nikjoo 2003). These Monte Carlo simulations describe several possible pathways for specific DNA damages and are useful for those of us who need to visualize processes.

To conclude this section on DNA damages produced by ionizing radiation in human tumor cells, it is our opinion that those that combine to produce cell killing by the β-mechanism are likely to be relatively simple lesions that are produced mainly by radiation "spurs" and are completely reparable. When low dose-rate therapies are employed, these "two-hit" lethal events never occur since there is adequate time between the events for repair enzymes to do their job. The radia-tion-induced DNA lesions that result in α-inactivation are produced by multiple ionizations in larger volumes at electron track-ends. These events are relatively rare and will have heterogeneous forms. These events are much more frequent when dose is delivered in Auger cascades or by the Bragg peaks of protons and heavy-charged particles. Several studies suggest that α-inactivation is greatest when DNA is in a condensed form and when it is associated with lipid elements in the nuclear membrane. As will be described in the next chapter, the majority of tumor cell killing by fractionated protocols of 1.8–3.0 Gy daily dose will be via the α-inactivation mechanism.

7 The Radiosensitivity of Tumor Cells *In Vitro* versus *In Vivo*

The intrinsic radiosensitivities of mammalian cells irradiated as monolayers or in suspension culture are compared to those of the same cells irradiated as multicellular spheroids or as solid tumors growing in rodent hosts. No systematic differences are observed that are not predicted from the *in vitro* measurement of tumor cell radiosensitivity, suggesting that these data will be reliable for use in TCP modeling. The intrinsic radiosensitivities of tumor clonogens released from human biopsy specimens are similar to the radiosensitivities of cell lines derived from similar tumor phenotypes. A table of $\bar{\alpha}$ and $\bar{\beta}_0$-inactivation parameters for several cancer pathologies suitable for input into TCP models is presented. These parameters should provide a useful starting point and should be updated as additional information about other biological parameters known to modulate radiosensitivity become known.

Up to this point, the killing of tumor cells by the α- and β_0-inactivation mechanisms of the linear-quadratic (LQ) model has been described and justified with experimental data for single cells irradiated *in vitro*, where biological, chemical and physical factors that impact outcome can be rigorously controlled. The question now arises as to whether we can be sure that these parameters for cell inactivation describe reliably the intrinsic radiosensitivity of tumor clonogens *in vivo*. Three different sets of experimental data that speak to this point are described in the following sequence: the irradiation of cells in multicellular spheroids, the irradiation of cells in solid rodent tumors and the *in vitro* assays of intrinsic radiosensitivity (SF_{2Gy}) of human cancer cells released from biopsy specimens.

7.1   The Radiosensitivity of Cells Irradiated in Multicellular Spheroids

Growing mammalian cells in slowly stirred suspensions resulted in aggregations of dividing cells when the stirring rate was slow. Under these growth conditions, cells continued to adhere to their division mates after cytokinesis and eventually produced spheres of many cells. These clumps of cells were called "spheroids" arising from single cells and could reach diameters of a least 1 mm (Durand and Sutherland 1972). When Chinese hamster cells in spheroids were irradiated, dispersed with enzymes and plated for colony-forming ability, they exhibited a greater resistance than when irradiated as single cells (Durand and Sutherland 1973a). Figure 7.1 shows survival curves for this effect from the original studies. Subsequent research indicated that a portion of this resistance could be due to some cells residing in a quiescent phase (G_0) of the growth cycle (such cells are more radioresistant than G_1-phase cells; see Chapter 3), while some portion of the effect was postulated to result from enhanced repair of sublethal damage (Durand and Sutherland 1973b). The data were not analyzed by the LQ equation but subsequent analyses indicate that the majority of the observed

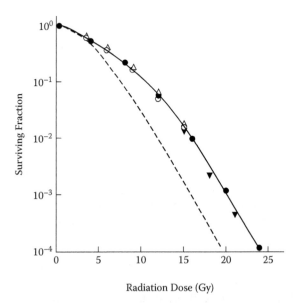

Radiation Dose (Gy)

Figure 7.1. The radiation sensitivity (surviving fraction) of Chinese hamster V79 cells irradiated in multicellular spheroids. The dashed line is their expected killing when irradiated as single cells. (Reprinted with permission from Durand, R. E. and Sutherland, R. M. 1973a. *Radiation Research* 56:513–527.)

radioresistance resulted from a lower α-inactivation mechanism. This spheroid technique has provided a wealth of information about cell growth compartments (proliferating and quiescent), the development of hypoxic microenvironments (limited by oxygen diffusion and its metabolic consumption) and repair processes in multicellular systems (Sutherland 1988). It has been exploited by several investigators to address various aspects of tumor biology (Culo, Yuhas and Ladman 1980). Mammalian cells irradiated as contact-inhibited monolayers or in spheroids are always more radioresistant than when irradiated as single cells.

7.2   The Radiosensitivity of Rodent Tumor Cells

Additional radiobiology information has been derived from studies performed with transplantable rodent tumor systems. An early tumor cell assay involved the irradiation of rodent cells *in vitro* and the determination of their tumor-producing efficiency (TCD_{50}) when injected back into immune compatible mice (Hewitt and Wilson 1961). This technique was later modified to score tumor cell colonies (usually about 10–40 colonies) in the spleens or lungs of mice that arose after the intravenous injection of tumor cells (irradiated and unirradiated) back into host animals (Till and McCulloch 1961; Hill and Bush 1969). This procedure greatly improved the "statistics" of this assay by essentially using one living animal as one "petri dish." Another technique for quantifying the effectiveness of different radiation procedures with rodent tumors involves measuring their dimensions with calipers and computing their volumes prior to and after treatment (Barendsen and Broerse 1969). Figure 7.2 shows the volumetric changes induced by various radiation doses to Dunning AT rat prostate tumors (a rapidly growing, anaplastic variant) and Dunning H tumors (a slowly growing, well-differentiated variant) (Thorndyke et al. 1985). The larger growth delay observed in H tumors treated with 35 Gy relative to that in AT tumors was attributed to its smaller growth fraction and its better oxygenation status. Other studies had measured the hypoxic fractions of the AT and H tumors to be ~18% and <1%, respectively (unpublished data). In fact, even after an acute dose of 50 Gy, the AT tumors continued to grow (by volume increase) for several days before the cell-killing effects of radiation were expressed. When the regrowth curves are extrapolated back to zero time, a dose of 50 Gy to the AT tumors and 25 Gy to the H tumors appears to have killed about three logarithms of clonogens. This rat tumor model provided useful information about tumor perfusion and hypoxia as measured by nuclear medicine imaging agents (Moore et al. 1992, 1993), tumor hypoxia measured by the Eppendorf pO_2 histograph (Yeh et al. 1995) and treatment response to chemotherapy, radiotherapy and photodynamic therapy (Mador et al. 1982; Thorndyke et al. 1985; Gonzalez et al. 1986). This growth-delay assay mimics most closely how oncologists assess tumor treatment response today; that is, they observe the volumetric response of treated tumor volumes (PTVs) by palpation (when possible) and/or by appropriate imaging procedures. In most cases, the tumor size is described by the length of its longest dimension observed in a diagnostic image and/or by maximal cross-sectional area. While these assays have been useful for understanding radiation effects *in vivo*, they cannot directly

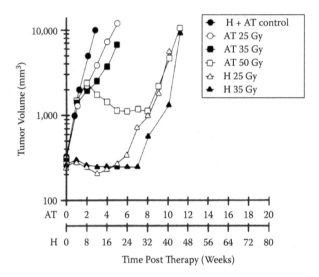

Figure 7.2. The volumes of R3327-AT (anaplastic) and R3327-H (well-differentiated) tumors irradiated at ~0.3 cm^3 with various doses of ^{137}Cs-γ-rays. There is a fourfold difference in doubling times of these tumors and their time response has been normalized so that the growth rates of unirradiated tumors are similar. (Adapted and reproduced with permission from Thorndyke, C. et al. 1985. *Journal of Urology* 134:191–198.)

produce the LQ parameters required for tumor response modeling. Pathological examination of irradiated tumor tissue specimens gave insights into the roles of proliferative cell death, edema and clonogen proliferation in these tumor volume responses (Thorndyke et al. 1985).

The rodent tumor assay that is most amenable to quantitative investigation is known as the *in vivo/in vitro* assay. It is accomplished with mouse and rat tumor cell lines that have been conditioned to grow in tissue culture. These are injected (usually subcutaneously or intramuscularly to restrict tumor spread) into syngeneic animals, allowed to grow into tumors of desired size, irradiated *in vivo*, resected from the sacrificed animals, dispersed and then plated for cell colony-forming assays *in vitro*. The survival curves generated with different radiation doses by this technique are amenable to analysis by the LQ model. Since the procedures associated with this technique are much more numerous and complex and performed over longer times than those with cells *in vitro*, it is likely that larger errors will be associated with the inactivation parameters they produce. As well, that the cell populations derived from tumors of similar size in different animals are identical is usually an unproven assumption. Nevertheless, this technique has become the workhorse of many rodent tumor biology studies over the past 30 years. Figure 7.3 shows survival data from *in vitro* colony assays of EMT-6 tumor cells irradiated as tumors *in vivo* (from Rockwell and Kallman as published in Hall 2000). Survival data for these cells when irradiated *in vitro* (as single cells) in air and hypoxia are also shown. It is apparent that the initial response of

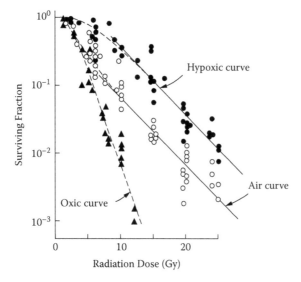

Figure 7.3. The surviving fractions of mouse EMT-6 tumor cells irradiated as tumors *in vivo*, resected, dispersed and plated for colony formation assays *in vitro*. The radiation sensitivity of aerobic and hypoxic EMT-6 tumor cells is also shown. (Reproduced with permission from Hall E.J. 2000. *Radiobiology for the Radiologist*, Lippincott Williams & Wilkins.)

the cells in tumors is similar to that of aerobic cells and that, at higher doses, the kinetics of killing conform to that of hypoxic cells. The downward displacement of the resistant tail of the tumor cell survival curve gives an estimate of the average hypoxic fraction (HF) present in the tumors at the time of radiation. The variation in survival data is much larger than that observed for single cell assays, indicative of the heterogeneity of the cells in the animal tumors. Though less precise, the LQ parameters derived from survival curves generated for cells irradiated as solid tumors *in vivo* are not significantly different from those generated with cells irradiated *in vitro*. This suggests that radiation inactivation parameters derived from cells irradiated *in vitro* can be trusted to predict the effects of radiation treatments on clonogens in solid tumors. It should be noted that tumor response *in vivo* can be influenced by several factors, including growth fraction, hypoxic fraction and immunological processes, that are not in play in petri dishes.

7.3   Appropriate Inactivation Parameters for Modeling Human Tumor Response

So what parameters of the LQ equation should be employed to model the response of different human cancers? Fertil and Malaise (1981) were first to suggest that the SF_{2Gy} of human tumor cell lines correlated with the clinical response of human cancers from which the lines were derived. Deacon, Peckham and Steel (1984) "culled" this list of human tumor cell lines and grouped their

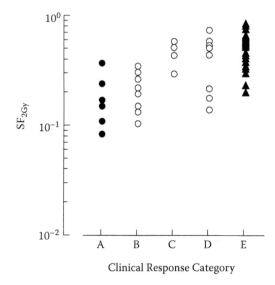

Figure 7.4. The surviving fraction after a 2 Gy dose (SF_{2Gy}) for 51 human tumor cells lines ordered according to their clinical response from most sensitive to most resistant (left to right). (Reproduced with permission from Deacon, J. et al. 1984. *Radiotherapy Oncology* 2:317–323.)

radiation responses (SF_{2Gy}) into five classes, from the most responsive to the most radioresistant (see Figure 7.4). But a large variation in the intrinsic radio-sensitivities of the tumor cell lines within each group is evident, with the most resistant in the most responsive class having a higher SF_{2Gy} than the most sensitive of the most radioresistant class. This indicates that a large variation in tumor cell radiosensitivity should be expected between cancers of similar pathology in different patients (cohorts that are used in investigative clinical protocols) for each class of tumor, and this should be accounted for in TCP modeling (see Chapter 10).

Figure 7.5 shows survival curves for seven human tumor cell lines that were generated with asynchronous populations in slowly stirred suspensions irradi-ated at low temperature. The left panel presents the data in the traditional semi-log plot and the right panel shows the same data plotted as $-[ln(SF)]/D$ versus dose (see Equation 2.3, Chapter 2). While small differences in the slopes ($\bar{\beta}_o$-mechanism) of the survival curves (right panel) are observed, the largest variation in tumor cell radiosensitivity is the intercept of the lines with zero dose (the $\bar{\alpha}$-mechanism), presumably related to their radiosensitivity in the interphase (see Chapter 3). Table 7.1 gives the best-fit values of and $\bar{\alpha}$ and $\bar{\beta}_o$ derived from such survival curves along with values for $\bar{\alpha}/\bar{\beta}_o$ and SF_{2Gy} for 10 human tumor cell lines irradiated as asynchronous populations at low temperature. It is obvious that the largest variation in tumor cell radiosensitivity results from differences in the $\bar{\alpha}$-mechanism (a 45-fold difference) and the $\bar{\alpha}$ and $\bar{\beta}_o$ ratios are certainly not constant (Peacock et al. 1992).

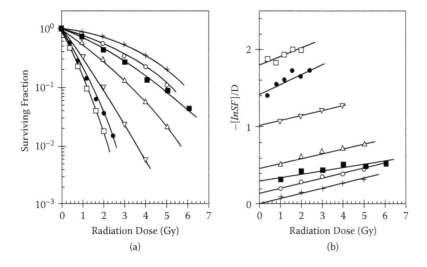

Figure 7.5. Survival curves for seven asynchronous populations of human tumor cell lines irradiated at low temperature with ^{137}Cs-γ rays. The data are plotted in the traditional format (left panel) and as $-[lnSF]/D$ versus dose (right panel). The different symbols are for HT-29 (plus), OVCAR10 (open circle), MCF-7 (closed square), A2780 (open triangle), HX-142 (inverted open triangle), HT-144 (closed circle) and Mo59J (open square).

Table 7.1. Parameters of *In Vitro* Intrinsic Radiosensitivity of Asynchronous Human Tumor Cell Lines Irradiated at ~5°C

Cell Line	Tumor Origin	$\bar{\alpha}$ (Gy^{-1})	$\bar{\beta}_o$ (Gy^{-2})	$\bar{\alpha}/\bar{\beta}_o$	SF_{2Gy}
HT-29	Colon	0.04	0.063	0.64	0.73
TSU	Prostate	0.06	0.048	1.24	0.70
OVCAR10	Ovary	0.16	0.052	3.10	0.58
PC-3	Prostate	0.24	0.068	3.55	0.48
DU-145	Prostate	0.31	0.048	6.46	0.48
MCF-7	Breast	0.38	0.026	14.6	0.43
A2780	Ovary	0.47	0.073	6.44	0.29
LnCap	Prostate	0.49	0.015	7.84	0.25
HT-144	Melanoma	1.43	0.130	11.0	0.03
Mo59J	Glioblastoma	1.80	0.096	18.8	0.01

Average $\bar{\beta}_o = 0.0581 \pm 0.0043$.

Much of the data on intrinsic tumor cell radiosensitivity is reported as the surviving fraction observed after a 2 Gy dose. If the average $\bar{\beta}_o$ value (0.0581 Gy^{-2}) derived from the data in Table 7.1 is assumed to be representative of the majority of human tumor cell lines, $\bar{\alpha}$-inactivation parameters can be computed from the SF_{2Gy} data with the following equation:

$$\bar{\alpha} = \left(-\ln SF_{2Gy} - 4\,\bar{\beta}_o\right)/2 \qquad (7.1)$$

Table 7.2 shows average values of $\bar{\alpha}$-inactivation parameters ± SD and the "fixed" $\bar{\beta}_o$-inactivation parameter for the five groups of tumors shown in Figure 7.4. Since there was no significant difference in the intrinsic radiosensitivities of groups A and B or in groups C and D, the data were combined to produce the average $\bar{\alpha}$-inactivation values for only three classes of human tumors. These are then reasonable values of intrinsic radiosensitivity for the various cancers represented by the tumor cell lines in Figure 7.4. The error estimates for these $\bar{\alpha}$-parameters should account for the variation expected from interpatient heterogeneity.

The intrinsic radiosensitivity of human cervical cancer cells was reported by West et al. (1993, 1997). These studies involved the enzymatic dispersal of clonogens from fresh tumor biopsy specimens, irradiating the cells *in vitro* and determining SF_{2Gy} by clonogenic assays. The mean ± SE value for $\bar{\alpha}$-inactivation from these data (using the average $\bar{\beta}_o$-inactivation described previously) was 0.35 ± 0.21 Gy^{-1} (see Table 7.2). The intrinsic radiosititivity (defined by α) of clonogens recovered from cervical cancer biopsies is similar to that of the cervical cancer cell lines (Fertil and Malaise 1981). Similar studies performed by Björk-Eriksson et al. (2000) with biopsies from head and neck carcinomas were analyzed in the same manner. A mean value of $\bar{\alpha} = 0.40 \pm 0.21$ G^{-1} (see Table 7.2) was obtained and correlates well with values for head and neck cancer cell lines (Fertil and Malaise 1981).

Another study that modeled prostate tumor response to brachytherapy and conformal radiotherapy (Nahum et al. 2003) used data pooled from three different laboratories that had reported the radiosensitivity of prostate cancer cell lines in terms of the LQ model (see Table 7.2). The mean value of $\bar{\alpha}$-inactivation was $0.26 = \pm 0.17$ Gy^{-1}, and the $\bar{\beta}_o$-inactivation was 0.0313 Gy^{-1}. For the data generated by Algan et al. (1996) at low temperature during irradiation (where sublethal damage repair was minimized), the mean $\bar{\beta}_o$-inactivation parameter for DU-145, PC-3 and TSU cells lines was 0.0552 Gy^{-1}. A lower β-parameter would be expected in the other data sets since the radiations were administered at 24°C and 37°C, where some repair of sublesions could occur. But since the $\bar{\beta}_o$-inactivation comprises a minor proportion of total tumor cell killing expected in patients treated with low dose fractions or brachytherapy, these differences in the β_o-parameters will not be critical when modeling tumor response for conventional protocols.

Table 7.2. Appropriate Intrinsic Radiosensitivity Parameters for Various Human Tumor Cell Lines (Figure 7.4) and Clonogens Released from Biopsy Specimens for TCP Modeling Input

Tumor pathology	α(Gy^{-1})	β_o(Gy^{-2})
Groups A and B: lymphoma, myeloma, neuroblastoma, medulloblastoma and SSLC	0.73 ± 0.23	0.0581
Groups C and D: breast, bladder, cervical carcinoma, pancreatic, colorectal, head and neck and squamous lung cancer	0.36 ± 0.25	0.0581
Group E: melanoma, osteosarcoma, glioblastoma, renal carcinoma and prostate cancer	0.26 ± 0.17	0.0581
Cervical carcinoma (West et al. 1993)	0.35 ± 0.21	0.0581
Head and neck carcinoma (Björk-Eriksson et al. 2000)	0.40 ± 0.21	0.0581
Prostate carcinoma (Algan et al. 1996)	0.26 ± 0.17	0.0313

The data in Table 7.2 are the most appropriate values of intrinsic radiosensitivity $(\bar{\alpha}$ and $\bar{\beta}_o)$ currently available for modeling the response of most human tumors treated by current fractionation schemes. A single 2 Gy fraction will kill 70%, 61% and 53% of aerobic tumor cells in groups A and B, groups C and D and group E, respectively, by $\bar{\alpha}$-inactivation alone. The supposedly informative parameter, α/β, for these different tumor types is 12.6, 6.2 and 4.5, respectively, and not ~10, as frequently cited. When hypofractionation of dose is contemplated with 10 Gy fractions, the proportion of tumor cell killing by the α-inactivation mechanism will be reduced to 56%, 38% and 31% for the three tumor groups, respectively. This should serve as a warning to those that propose to optimize treatment plans by prescribing fractions of significantly higher dose. The $\bar{\beta}_o$-inactivation mechanism will become more dominant as the fraction size increases (as described in Chapter 5) and its characteristics are significantly different from those of the cell-killing mechanism that dominates in current radiotherapy.

Figure 7.6 shows the expected survival curves for the three groups of tumor cells of Table 7.2 when irradiated at 4°C, at room temperature and at 37°C at a dose-rate of 0.3 Gy/min. When survival curves are generated with cells in suspension chambers (Figure 2.2, Chapter 2), the beam-off times required for obtaining cell samples after specific dose fractions should be factored into the overall treatment time and "effective" dose-rate. While the difference between cell killing for radiation administered at low temperature (no repair) and room temperature at this dose-rate is small, the increased survival observed when cells are irradiated at 37°C is significant. This will also be relevant for modeling clinical fractions that are delivered in multiple ports, each day. The overall treatment times will

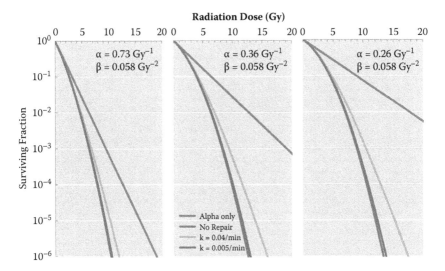

Figure 7.6. Survival curves generated with Equation (5.2) (in Chapter 5) for tumor cells with intrinsic radiosensitivities of those in groups A and B, groups C and D and group E irradiated at low temperature (no repair of sublesions) and at room temperature and 37°C.

dictate the extent of sublethal damage repair during the exposures. The repair of sublesions during the radiation exposures at 37°C will impact tumor cell killing the most for cells with the lowest $\bar{\alpha}$-parameters. The survival curves of Figure 7.6 were generated with Equation (5.2) (in Chapter 5). And if survival data generated with cells at 37°C are expressed with only two parameters, α and β, the resultant β will be lower, and consequently α will be larger than the true inactivation parameters.

As will be discussed in Chapter 9, the α-inactivation parameters of normal tissue fibroblasts fall within the range of radiosensitivity of the most resistant tumor cells (group E). The radiosensitivities of stem cells in bone marrow and intestinal mucosa are larger than those of fibroblasts—more like those of the tumor cells of groups A and B. The success of current radiotherapy prescriptions empirically optimized over several years might well have exploited the fact that most tumor clonogens (some proliferating) exhibit a higher α-inactivation parameter relative to those in quiescent normal tissues. The most appropriate $\bar{\alpha}$ and $\bar{\beta}_o$-inactivation parameters for both tumor and normal tissue cells should be updated as more experimental data become available.

8 Modern Radiobiology and the LQ Equation

Current radiobiology research appears to have distanced itself from the mechanisms associated with α- and β_o-inactivation. In many studies, little regard is given to the fact that two independent mechanisms are in play and that their properties are qualitatively distinct. The dominant mechanisms of cell killing in populations irradiated with doses of 2 and 20 Gy are very different. Much of the current search for molecular factors that might inform about cell radiosensitivity reads like a "molecular taxonomy." That is, there is a desire to identify individual or groups of molecular expressions that will correlate with intrinsic radiosensitivity. Of that research, the studies that improve our understanding of repair mechanisms of radiation-induced lesions in cellular DNA are the most relevant to radiotherapy. They should inform about the β_o-mechanism, particularly in normal tissues, where radiation damage repair between dose fractions is essential. Low-dose hypersensitivity has been researched extensively; how it might relate to linear-quadratic (LQ) radiobiology and radiotherapy prescription of fractioned dose is discussed. The radiation bystander effect consumes a large portion of current radiobiology research funding but its importance for tumor cell eradication *in vivo* is not particularly clear. The reader is directed to other radiobiology textbooks that describe several other phenomena (cell cycle regulation, immunology, etc.) that might play a role in tumor cell killing by radiation. To have a significant benefit for clinical practice and tumor response modeling, radiobiology studies should be related to the two independent mechanisms of cell killing described by the LQ model.

Much of modern radiobiology research has gravitated toward the identification of molecular expressions that might relate to cellular radiation responses. The modern tools of molecular biology are almost a prerequisite for obtaining research grants in this field and are held in high regard for their potential ability to advance our understanding of the cell processes involved in radiotherapy. This trend should not be overturned, but it is our opinion that the "radiation" component of radiobiology must be incorporated into these studies if the aims of this research are to succeed. The basic mechanisms of cell killing in play after exposures to 20 Gy are significantly different from those in cells that are treated with 2 Gy fractions. And this difference becomes exaggerated when 2 Gy fractions are delivered each day over several weeks. This fact is currently insufficiently emphasized and the radiation dose of choice in most experiments is that which gives measurable changes in the protein, RNA, DNA or lipid expressions or functions that are under investigation. That is not to say that these experiments will have little or no meaning, but only that they should take into account what is already known and, in particular, what is known from the LQ model.

8.1   Molecular Biology Factors of α- and β-Inactivation

The single-hit mechanism of human tumor cell inactivation has not been well researched. Reasonable estimates of $\bar{\alpha}$-inactivation parameters can be obtained from human tumor cell lines and tumor biopsy specimens. And since the single-hit mechanism produces the majority of cell killing in current fractionated treatment protocols that lead to cancer cures, its molecular basis should be better defined so that it can be optimized and manipulated for improved treatment. Its most important characteristic is that it exhibits no potential for repair, at least in split-dose and dose-rate experiments that demonstrate complete sublethal damage repair of the $\bar{\beta}_o$-mechanism (Chapman 2003). The maximal rate of single-hit inactivation measured for mammalian cells exposed to low-LET radiation is 2–5 Gy^{-1}; these values are exhibited by several DNA repair-deficient cells lines and cells at mitosis (Chapman et al. 1999). On the other hand, $\bar{\alpha}$-inactivation rates as low as 0.02–0.05 Gy^{-1} have been reported for some resistant human tumor cell lines—two orders of magnitude lower than the most sensitive cells. It is the increase in single-hit killing of the LQ model that describes the superior effectiveness of high-LET radiations (see Chapter 4) and the major differences in radiosensitivities between human tumor cell lines (see Chapter 7). More research is required to better understand the ionizing events, the molecular targets and the molecular damages that produce radiation killing by this mechanism; biophysical and molecular biology research could be extremely informative.

Since mitotic cells exhibit this maximal radiosensitivity, DNA in compacted form might constitute an ultrasensitive molecular target (Chapman et al. 1999). And if electron track-ends are the ionizing entities that produce the associated molecular lesions, the damaging events are likely to be very local, highly variable and relatively few and far between (Chapman 1980). This suggestion is consistent with the observations of Cole and colleagues, who irradiated cells with low-energy electrons and found that when they stopped in the vicinity of the nuclear membrane, cell killing was by single-hit kinetics, exhibited no oxygen effect and showed no potential for repair (Cole et al. 1974, 1975; Tobleman and Cole 1974). The molecular target for cell inactivation by the single-hit mechanism was postulated to be some complex consisting of compacted DNA, protein and lipid that resided near the nuclear membrane (Lange, Cole and Ostashevesky 1993). Regions of condensed chromatin adjacent to the nuclear membrane were observed in interphase repair-deficient cells by electron microscopy and could present a larger cross section for electron track-end events (Chapman et al. 1999). And since the oxygen diffusion kinetics from the radiation targets in tumor cells at mitosis are similar to those from cellular lipids, the molecular makeup of their targets for single-hit inactivation could include lipid derived from the nuclear membrane (Stobbe, Park and Chapman 2002).

The process of chromatin condensation in mammalian cells has been extensively investigated over the past 20–30 years and is described schematically in Figure 6.9. Histone processing by acetylation, methylation and phosphorylation occurs throughout the cell cycle and plays an important role in DNA compaction at the time of mitosis, producing the chromosomes that segregate to daughter cells at cytokinesis. The phosphorylation of histone 3 is a "molecular signature" for chromatin in mitotic cells (Gurley et al. 1978) and was observed in HT-29 tumor cells that had been treated with compounds known to produce premature chromatin condensation (Biade et al. 2001; Price et al. 2004). Phosphorylated H3 and/or other histone modulations unique to condensed chromatin should be investigated as a rapid assay for predicting tumor cell radiation sensitivity by the α-mechanism. This ought to be an important area of current cell and molecular biology research as it will impact significantly our understanding of the intrinsic radiosensitivity of tumor clonogens.

Several molecular factors that relate to the β_0-inactivation mechanism have been described. The sublethal lesions associated with this mechanism exhibit complete repair when radiation is administered at a low dose-rate (see Chapter 5). Presumably, all of the simple molecular sublesions (base damages, single-strand breaks [SSBs], double-strand breaks [DSBs], etc.) involved in this cell inactivation process are repairable. The description of several pathways for the repair of DNA base damages, SSBs and DSBs constitutes a large literature that was recently reviewed (Olive 2000; Branze and Folani 2008; Clancy 2008). Most of these DNA damages in irradiated cells result from "spurs" that constitute the greater part of the dose delivered to human tumor and normal tissue cells during radiotherapy. Between tumor dose fractions, these repair processes restore cell radiation sensitivity to its preirradiation state (see Equation 5.1, Chapter 5). This is also an essential process in the normal tissue cells (see Chapter 9) that are exposed during clinical treatments. Since these DNA repair pathways exist for the maintenance of DNA integrity during the normal lifetime of cells in body tissues, the majority of non-mutated cells will express them at high proficiency. The induction and repair of DNA strand breaks appear to play a role in chromatin modulation throughout the cell cycle (Stobbe et al. 2002) and can be expected to operate in the tissues of most patients presenting for radiotherapy treatment. In the case of patients with atexia telangiectasia, xeroderma pigmentosum, Bloom's syndrome and some other diseases, some component of their DNA repair enzymes is mutated, which can cause radiation hypersensitivity (Pollard and Gatti 2009). These phenotypes constitute a very small fraction of the patients coming for radiotherapy and can usually be identified prior to treatment so that prescriptions can be adjusted.

While DNA repair pathways, cell cycle checkpoints and cell signaling pathways in irradiated cells have been extensively investigated and characterized (Hall and Giacca 2006), this information is not routinely used to stratify patients for treatment. Since most of these molecular characterizations relate to cell killing by the β_o-mechanism, their impact on tumor response by current fractionation schemes will be relatively small. However, they will play a much greater role in the expression of normal tissue complications. If radiation oncology gravitates to hypofractionated treatments, tumor response will become more dependent upon the β_o-inactivation mechanism and this research will become more important for understanding tumor responses.

There is currently broad interest in the screening of human cancer cells for molecular expressions that relate to cellular radiosensitivity. Our description of this area of research as "molecular taxonomy" was meant to remind researchers of what has gone before in other fields of science. Just because studies like this can now be accomplished does not make them excellent research. And just because cancer granting agencies are enamored by molecular biology procedures does not negate the fact that excellent research should be hypothesis based and, hopefully, derived from a prior foundation of understanding. The "sifting" of associations from arrays of DNA, RNA and protein expressions in cells will hopefully identify some cellular processes that are most important in the expression of intrinsic tumor cell radiosensitivity. These studies will best be performed if the known differences in the basic mechanisms of α- and β_o-inactivation are at the forefront in the analyses. Modern molecular radiobiology should strive to relate its findings to the kinetics of quantitative cellular responses to radiation that have been defined by past experimentation.

8.2   Low Dose Hypersensitivity (LDH)

Two research tools that have facilitated a better definition of the effects of low doses of radiation on mammalian cells are the DMIPS (dynamic microscope image processing scanner system) and fluorescence-activated cell sorting (FACS) systems (Jaggi and Palcic 1985; Palcic et al. 1988; Wouters, Sy and Skarsgard 1996; Joiner et al. 2001). These enable the observation of individual cells after irradiation to assess their progression to colony formation or to cell death. At doses below 0.5 Gy, some studies have shown an initial radiosensitivity that is greater than that measured at larger doses. Thus, the term "low-dose hypersensitivity" (LDH) was invoked to describe this radiation response, and the cell line T98G (human glioma) showed the largest LDH effect (Short et al. 1999). Precise radiosensitivity measurements for the first 5% or less of cell killing are difficult to achieve by normal cell plating procedures. The relatively flat response observed between 0.2 and 1.0 Gy in T98G cells was interpreted as increased

DNA repair capability in surviving cells that was radiation induced. There could be alternative explanations for this effect. The majority of these studies were performed with asynchronous populations of mammalian cells that are known to exhibit different radiosensitivities throughout their growth cycles (see Chapter 3). Mitotic HT-29 colon cancer cells exhibit an α-inactivation parameter of 1.8 Gy^{-1} compared to a value of 0.05 Gy^{-1} for the same cells in interphase, almost a 40-fold difference. A smaller difference of about 12-fold was observed between the initial and ultimate killing rates of T96G glioma cells (Short et al. 1999). If 5% of an asynchronous population of HT-29 cells was mitotic (or hypersensitive by virtue of some other mechanism) at the time of irradiation, a biphasic survival curve at very low dose would be expected and should conform to Equation (3.1) in Chapter 3. Thus, heterogeneity of radiation response in asynchronous populations of tumor cells might explain some low-dose hypersensitivity observations. Skarsgard, Wilson and Durand (1993) performed similar studies with synchronized Chinese hamster cells irradiated in the G$_1$- and S-phases and found the low-dose hypersensitivity effect to be greatly diminished. For most cell lines that have been investigated in this manner, the initial hypersensitivity accounts for 5% or less of the cell inactivation (Marples and Collis 2008).

If an increase in DNA repair capacity is the explanation for this low-dose hypersensitivity effect, temperature could be used to determine if metabolic processes are required or whether it is related to some biophysical rearrangement (from sensitive to resistant structure) of the chromatin target. Both possibilities are intriguing and are amenable to experimental investigation, but the procedures involved in the DMIP and FACS techniques are complex and costly to perform. In the end, it is unlikely that a clearer knowledge of the underlying mechanism(s) of these experimental observations will impact significantly our current radiotherapy procedures. Clinical experience has demonstrated that daily doses of 1.8–3 Gy will kill at least 50% of the clonogens in most tumors, and several days or weeks of such treatments are required to eradicate the threat of further tumor growth (see Chapter 10). Hyperfractionation protocols have employed two to three smaller dose fractions on each day and some achieve an acceptable tumor response. But the administration of 0.2 Gy fractions several times each day would be labor intensive and would only be tested if the promise of clinical benefit was exceptional. It is not yet clear if cells recover their initial hypersensitivity to radiation between multiple fractions and over what times. Consequently, our opinion is that while low-dose hypersensitivity has been an intriguing phenomenon to investigate in the laboratory, it is unlikely to yield significant benefits for practical radiotherapy. Smith et al. (1999) observed little or no excess cell killing when 6 Gy was delivered in 0.3 Gy fractions compared with 2 Gy fractions.

8.3   Bystander Effects

That radiation can have effects by both direct and indirect mechanisms has been part of the teaching of radiobiology for many years. Both direct ionizations within the chromatin in the cell nucleus and in the adjacent water can lead to cell killing as measured by loss of proliferative capacity (see Chapter 6).

Under aerobic conditions, the indirect effects of OH' are responsible for at least 65% of the cell killing produced by x-rays and heavy-charged particles. For Auger electrons produced by isotope disintegrations in cell DNA, the proportion of cell killing by the indirect effects of OH' is ~90% (Walicka, Wilson and Durand 1998a). These indirect effects produced within individual cells have never been described as bystander effects, and the diffusion distance of the involved toxic radicals is only 2–3-nm.

While the DNA (chromatin) in mammalian cells was shown to be the major target for radiation damages that lead to their death in tumor cells, the treatment of cancer patients with ionizing radiation will expose tissues outside the planned treatment volumes (PTVs) with dose that could have detrimental consequences (normal tissue complications). Studies with charged particle beams and isotope sources (see Chapter 6) suggested that these other effects were much less important, at least for cell killing as measured by loss of proliferative capacity.

There has been a recent renaissance in research to investigate the effects that irradiated cells might have on adjacent or nearby cells that are not exposed. These studies are now described as investigations of "bystander effect." Many are accomplished with single-beam or single-particle radiation sources, with improved microscopic techniques for localizing dose to within micrometer and smaller regions within cells and with compartmentalized culture dishes that allow irradiated cells to "communicate" with unirradiated cells via a common growth medium (which may or may not have been irradiated). While presentations on bystander effects command a large portion of the program at current meetings of radiation researchers, no procedures have yet been identified that could significantly improve current radiotherapy practice.

The research on multicell spheroids described in Chapter 7 indicated that cells irradiated in close proximity generally exhibit a lower radiosensitivity than when irradiated as single cells. Those studies suggest that free radicals and/or other radiation-induced diffusible factors produced outside individual clonogens in tumors will not increase the cell killing and therefore will not translate into improved tumor response. It is possible that some component of the radiation killing of attached or suspended cells is from free radicals generated in the growth medium and that the "packing" of cells together in spheroids, tumors and normal tissues reduces that effect. Since the indirect effects of water radicals result from extremely short diffusion distances (2–3-nm), they are unlikely to be able to diffuse into cells and through the cytoplasms to damage cell chromatin. That is not to suggest that cell-to-cell communication processes play no role in tissue response to radiation but, for most toxins (free radicals and others) generated by ionizing radiation, their reactions with other molecules are extremely rapid and occur within a few nanometers of where they are produced. This is certainly the case for OH' and e^{-aq} (see Chapter 6). H_2O_2 reacts with cellular molecules at much lower rates, can be removed from cells by enzyme reactions and could play a role in some bystander effects. But the concentrations of H_2O_2 produced within cells by 1.8–3 Gy dose fractions is very small and is efficiently eliminated by peroxidases and superoxide dismutase, in particular. Current research is directed to identifying other radiation-induced products in water or in irradiated cells that can diffuse to and produce deleterious effects in unirradiated cells. To date, the cellular responses that are measured in bystander effect research, including cell killing, cell mutation and various molecular expressions. have not identified strategies for improving the killing of tumor clonogens that will translate into improved clinical response. Bystander effect research may be more relevant for radiation-induced normal tissue damage that is described in the next chapter.

This brief review of some modern radiobiology research is mainly for the purpose of determining whether or not it has produced results that strongly impact on our understanding of cell inactivation by the α- and β_0-mechanisms of the LQ model and therefore how it might impact tumor response modeling. Unfortunately, for the greater part of molecular biology, LDH and bystander effect research, not enough consideration has been given to relating it to the α- and β_0-inactivation mechanisms. As the above fields of research continue to be developed, they should incorporate current knowledge of cell killing by these two independent mechanisms and how this research might lead to improved clinical protocols.

9 Normal Tissue Radiobiology

Cell lines of rodent and human fibroblasts from lung, ovary, kidney and other tissues along with lymphocytes and bone marrow stem cells have been investigated using the classical techniques of *in vitro* tissue culture. These studies produced useful values of $\bar{\alpha}$- and $\bar{\beta}_o$-inactivation parameters for normal tissue clonogens. In those cases where normal tissue damage results from the inactivation of critical stem cells, it is likely that these data will be useful for predicting associated radiation complications. The α-inactivation parameter of normal tissue clonogens is usually lower than that of tumor cell lines derived from the same tissues. This may result from chromatin rearrangements in the tumor cells and from a preponderance of quiescent cells in the normal tissues. That these quantitative parameters can predict for the intrinsic radiosensitivities of normal tissues *in vivo* has yet to be demonstrated. Some radiation complications arising after clinical radiotherapy are likely to be due to damage produced in stromal components (vasculature and structural) of the tissues and not from damaged parenchymal cells. The majority of our understanding of normal tissue tolerance comes from animal studies and clinical experience. These data have recently been updated and are reported in the QUANTEC study. It is apparent that our understanding of radiation-induced normal tissue complications of concern to radiation oncologists is "less quantitative" than our understanding of tumor cell killing. However, normal-tissue fractionation radiosensitivity can indicate appropriate α/β values for NTCP (normal tissue complication probability) modeling, which has become an important goal for several research programs.

9.1   Information Derived from *In Vitro* Studies of Normal Tissue Cell Lines

Cell lines derived from various normal tissues can be manipulated in tissue cultures to produce survival curves amenable to analysis by the linear-quadratic (LQ) equation. The data in Figures 2.1 (Chapter 2), 3.2 (Chapter 3), 4.2 (Chapter 4) and 5.4 (Chapter 5) were generated with fibroblasts derived from Chinese hamster lung tissue. The data from the Lawrence Berkeley Laboratory shown in Figures 4.5 and 4.6 were generated with T-1 cells derived from human kidney tissue. Since there are no systematic differences in the radiosensitivities of tumor stem cells derived from biopsies and cell lines established from like tumors (as discussed in Chapter 7), it is reasonable to assume that the stem cells of normal tissues *in vivo* and cell lines derived from those tissues will also express similar radiosensitivities. It is from such *in vitro* data that the following conclusions are based.

Firstly, stem cells from normal tissues usually exhibit smaller α-inactivation coefficients than tumor lines derived from like tissues. And since the β_o-inactivation coefficient does not vary widely between all the cell lines investigated, it will be the higher single-hit killing of tumor clonogens relative to that of the normal tissue stem cells that translates into the positive responses produced by fractionated radiotherapy protocols. This difference will increase as the fraction size is decreased but the daily fraction size must be large enough to produce inactivation of about 50% or greater. Proliferative death appears to be the most important mechanism for inactivating the stem cells of epithelia.

Secondly, *in vitro* characterization of stem cells from lymphatic tissues and bone marrow (progenitors of most blood cells) exhibits a relatively high radiosensitivity (mainly by α-inactivation) with death occurring mainly by apoptotic pathways.

Thirdly, while these broad rules apply to those normal tissues whose cell lines have been characterized, cell lines of other normal tissues of interest have yet to be established. Consequently, most of our understanding of normal tissue radiosensitivity is derived from studies of various animal tissues and from clinical observations.

Fourthly, a general rule that emerges from these studies is that normal tissues that are undergoing rapid proliferation to maintain homeostatic levels of functional cells (bone marrow, epithelia of the stomach and gut, etc.) are the most sensitive to ionizing radiations. As described in Chapter 3, proliferating mammalian cells are more sensitive to radiation inactivation than non-dividing (quiescent) cells. Cells in tissues that are not undergoing rapid division (muscle, liver, brain, etc.) will consequently exhibit a significantly lower radiosensitivity. This effect is consistent with the observation that quiescent cells *in vitro* exhibit a smaller α-inactivation parameter than the same cells in the G_1-phase of the growth cycle (see Chapter 3). And the majority of quiescent epithelial cells in most normal tissues will be differentiated and not amenable to immediate proliferative cell death until stimulated to divide.

Radiation effects on normal tissues have been divided into those that occur within a few weeks (early effects) of radiation exposures and those that occur at later times (late effects), usually months or years (Steel 2007b; Stewart and Dörr 2009). In general, the early effects can be attributed to stem cell inactivation in the specific tissues and exhibit α-inactivation parameters somewhat lower than those of proliferating tumor cells. Consequently, the generalization that tumors and early radiation complications in normal tissue have a similar α/β ratio of ~10 was accepted. But since many stem cells will be in a quiescent state at the time of irradiation, their radiosensitivity will be lower than that of tumors of the same origin. This relative radioresistance is likely to be a major factor for the current success of fractionated therapy. For tissues that contain a large stem cell population that are in a proliferative state, a high radiosensitivity can be expected. The late effects of radiation in normal tissues can potentially result from both the depletion of stem cells and from radiation damage to the stromal elements of the tissue. Radiation damages to tissue vasculature, in particular, could play an active role in some of the late complications observed in some normal tissues (Park et al. 2012). And how these complications are expected to change with the higher dose fractions of hypofractionation is critical for these novel therapies. Since most normal tissues occur in repeated subunits of functional tissue, their "architecture" plays an import role in the expression of their radiation damage (see following). The field of radiation damages in normal tissues has recently been reviewed and related to the LQ model (Stewart and Dörr 2009).

9.2 Therapeutic Ratio

Curative radiotherapy invariably involves delivering high radiation doses to the normal tissues adjacent to the tumor. In a very general sense it is the *tolerance* of normal tissues to irradiation that determines the likelihood of tumor control; were this tolerance infinite, then a sufficiently high dose for 100% tumor control could always be delivered. Figure 9.1 illustrates this "competition" between tumor

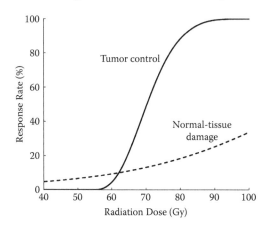

Figure 9.1. The dependence of clinical outcome (tumor control and normal-tissue damage) on radiation dose; the case illustrated here is for three-dimensional conformal radiotherapy, delivered in 30 daily fractions, for a non-small-cell (NSC) lung tumor (≈800 cc), and the "complication" is grade-2 radiation pneumonitis. The TCP and NTCP curves were generated by the BioSuite software (Uzan, J. and Nahum, A. E. 2012. *British Journal of Radiology* 85:1279–1286).

control and normal-tissue damage. Standard radiotherapy protocols specify total doses of around 60 Gy (with a fraction size of 2.0 Gy) for non-small-cell (NSC) lung tumors. For the example in this figure, it would be very unlikely to achieve tumor control: a dose ≈80 Gy would be required for a 90% chance of tumor elimination. However, this high dose would result in a ~20% chance of grade-2 radiation pneumonitis, which might be judged to be unacceptable.

Radiation oncologists have several tools at their disposal to increase the separation between the tumor control and normal-tissue damage curves i.e., to improve the TR [therapeutic ratio], which, though often not precisely defined, can be thought of as the probability of local control [or tumor control probability] for a given complication risk. One of the most important of these tools is *fractionation* (see next section); delivering a given total dose in either a large or a small number of fractions generally has a major effect on the TR. In the example illustrated in the figure the number of fractions is fixed (at 20) and therefore higher doses mean larger fraction sizes. There is also the option of delivering more than one fraction per day in order to decrease the overall treatment time, which can be advantageous for tumors with a high proliferation rate, such as the lung tumor featured in Figure 9.1 (Saunders et al. 1999).

Another key tool is beam-shaping, routinely done today by selecting the positions of the *leaves* of the multileaf collimator (MLC) so that the beam shapes correspond to the projection of the edge of the planned treatment volume (PTV) seen from the various beam directions (Mayles and Williams 2007); this "customized" beam collimation reduces the volume of normal tissue irradiated (or "increases normal-tissue sparing"). Modulation of the beam intensity, or intensity-modulated radiation therapy (IMRT) (Webb 2007), is a further step in sophistication, which may further decrease the risk of normal-tissue damage by *redistributing* the dose in the organs at risk.

Another important degree of freedom is beam modality: megavoltage x-rays of energies ranging from 4 to 20 MV, 4 to 20 MeV electrons (or a combination of these modalities) or possibly 250 MeV protons, moderated so that the Bragg peak is "spread out" over the tumor volume. For tumors in certain anatomical locations (prostate, breast, cervix, rectum) brachytherapy is another option.

The degree to which the preceding strategies succeed in reducing the probability of normal-tissue damage (i.e., in shifting the NTCP curve to the right) will depend on the precise nature of the complication in question. Put another way, in order to select an optimal treatment strategy it is desirable to know, in as much detail as possible, how the probability of damage (of a given type) depends on the magnitude and distribution of the dose (and number of fractions) in the normal tissue in question. The ideas involved in such dose-complication relationships will be outlined in subsequent sections.

9.3　Fractionation

Fractionation is key to the success of external-beam radiation therapy (Thames and Hendry 1987; Steel 2007a). Coutard (1929) established "standard fractionation" for treating head and neck cancer; a large number of small doses of radiation (typically 2 Gy), delivered each weekday, yielded a high "therapeutic ratio" (i.e., the best chances of local tumor control for acceptably low complication rates). However, it was not until some 50 years later, after the establishment of the LQ model, that fractionation was put on a firm radiobiological basis. The application of the LQ model to fractionated dose delivery proceeds as follows: for n equal fractions of dose d, assuming that sublethal lesion repair is completed in the interfraction interval (cf. Fowler 1989), the surviving fraction SF is given by

$$SF = \left[\exp\left(-\alpha d - \beta d^2\right)\right]^n$$
$$= \exp\left(-\alpha n d - \beta n d^2\right)$$

(9.1)

In terms of the total dose $D (= nd)$ this becomes

$$SF = \exp\left(-\alpha D - \beta d D\right)$$
$$= \exp\left[-\alpha D\left(1 + \frac{d}{\alpha/\beta}\right)\right]$$

(9.2)

where we recognize D $[1 + d/(\alpha/\beta)]$ as the *biologically effective dose, or BED,* defined as the dose delivered in an infinite number of tiny fractions that is radiobiologically equivalent to the regimen under study (i.e., n fractions of size d) (Fowler 1989). The term $[1 + d/(\alpha/\beta)]$, sometimes called *relative effectiveness, RE* (e.g., Steel 2007a), tends to unity as either the fraction size d tends to zero or (α/β) tends to infinity, which is consistent with the above definition of *BED*.

The concept of *iso*-effect has proven to be very useful in analyzing the effect of different fractionation regimens. Thus, one might want to know how the total dose D should change as the number of fractions (and hence the dose per fraction d) is changed for the *same* biological effect, expressed mathematically as a constant SF. Equation (9.2) can be rearranged as

$$D = \frac{(-1/\alpha)\log_e SF}{\left[1 + d/(\alpha/\beta)\right]}$$

(9.3a)

where, for the *iso*-effect condition for any normal tissue (or tumor), the numerator is a constant, denoted here by K:

$$D = \frac{K}{\left[1 + d/(\alpha/\beta)\right]}$$

(9.3b)

Figure 9.2 shows how total dose varies with fraction size d for different values of (α/β); for the curves in the figure, a value of 0.3 Gy^{-1} has been assumed for α, which, for 9 logs of cell kill, gives $K = 69.08$ Gy. From this figure it can be seen

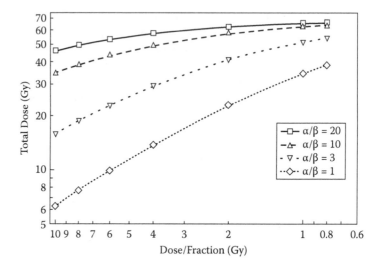

Figure 9.2. The variation of log (total dose *D*) with log (fraction size *d*) for *iso*-effect for different values of (α/β); Equation (9.3b) has been evaluated for $K = 69.08$ Gy corresponding to a surviving fraction of 10^{-9} and $\alpha = 0.3$ Gy^{-1}.

that the curves of total dose versus fraction size (which ranges here from 0.8 to 10 Gy) for *iso*-effect are almost "flat" for *high* values of (α/β) but relatively steep for low (α/β).

From fractionation studies on mice, Thames et al. (1982) demonstrated a systematic difference between *early* and *late*-responding normal tissues. Figure 9.3, from that paper, has attained "classic" status. By reference to Figure 9.2, where total dose is plotted against dose/fraction in a very similar manner, we can identify the curves in Figure 9.3 with a *shallow* gradient with *high* α/β and vice versa. Thus, for skin, *late* effects are associated with *low* α/β and *acute* (i.e., early) effects with *high* α/β. All the curves in Figure 9.3 are for studies with a short overall time, thus minimizing any effects of cell proliferation. A mixture of animal experiments, such as the one above, and empirical observations indicated that "late" normal-tissue (NT) damage was characterized by an α/β ratio of around 3 Gy (though with quite wide variations), whereas the corresponding curves for tumor *iso*-effect were best described by $\alpha/\beta \approx 10$ Gy (Thames et al. 1982). It follows that for a given level of tumor control (assuming negligible clonogen proliferation) the effects on normal tissues will continually *decrease* as the number of fractions *increases* and hence the fraction size *decreases*. Figure 9.4 illustrates this relationship in terms of biological effective dose for $\alpha/\beta = 3$ Gy, $BED_{\alpha/\beta=3}$, for a constant $BED_{\alpha/\beta=10}$.

The whole rationale for the use of a large number of small fractions is essentially represented by the curve in Figure 9.4, which was derived assuming that the α/β for late normal-tissue damage (= 3 Gy) is significantly lower than that for tumor clonogens (= 10 Gy). If this is not the case, then fractionation will not improve the therapeutic ratio.

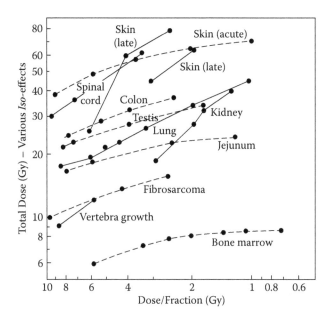

Figure 9.3. The dependence of total dose D on dose per fraction d for normal-tissue *iso*-effect in mice, for widely differing fractionation regimens; note that the d scale is reversed and that both axes are logarithmic, as in Figure 9.2. (From Thames et al. 1982; redrawn in Steel 2007a).

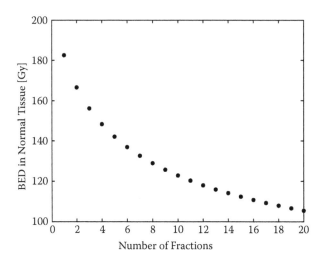

Figure 9.4. The dependence of *BED* on the number of fractions evaluated for $\alpha/\beta = 3$ Gy (see Equation 9.2), representing normal tissue damage, where the total dose has been determined for constant *BED* with $\alpha/\beta = 10$ Gy, representing tumor cells. Zero cell proliferation has been assumed.

In a radiotherapy clinic it is frequently desired to convert one fractionation scheme to another such that the two regimens are *iso*-effective with respect to a particular end-point, generally either tumor control or (late) normal-tissue effect. This involves equating the surviving fractions for the two regimens. If the current or *reference* regimen has total dose D_{ref} and fraction size d_{ref}, and the new regimen D_{new} and fraction size d_{new}, then for *iso*-effect, according to Equation (9.2), we must have

$$\alpha D_{new}\left(1+\frac{d_{new}}{(\alpha/\beta)_{str}}\right)=\alpha D_{ref}\left(1+\frac{d_{ref}}{(\alpha/\beta)_{str}}\right)$$

where the subscript "str" (i.e., structure) refers to either normal-tissue or tumor, depending on which *iso*-effect is required. As α is the same on both sides, the preceding expression is equivalent to equating *BED*s. The ratio of the "new" to the "reference" total doses is therefore

$$\frac{D_{new}}{D_{ref}}=\left(1+\frac{d_{ref}}{(\alpha/\beta)_{str}}\right)\bigg/\left(1+\frac{d_{new}}{(\alpha/\beta)_{str}}\right) \qquad (9.4)$$

This can be rearranged to yield

$$\frac{D_{new}}{D_{ref}}=\frac{(\alpha/\beta)_{str}+d_{ref}}{(\alpha/\beta)_{str}+d_{new}} \qquad (9.5)$$

which many readers will recognize as the "Withers *iso*-effect formula," (WIF) (Withers, Thames and Peters 1983). This formula is the standard method to calculate the change in the total dose when a change in the fraction size is contemplated.

Alternatively, the preceding formula can be expressed in terms of the change in the number of fractions, from n_{ref} to n_{new}, resulting in

$$\frac{n_{new}d_{new}}{n_{new}d_{ref}}=\frac{(\alpha/\beta)_{struc}+d_{ref}}{(\alpha/\beta)_{struc}+d_{new}}$$

which can be expressed as a quadratic equation in the unknown variable d_{new} with the positive root:

$$d_{new}=\frac{1}{2}(\alpha/\beta)_{str}\left\{-1+\sqrt{1+4\frac{n_{ref}}{n_{new}}\frac{d_{ref}}{(\alpha/\beta)_{str}}\left[1+\frac{d_{ref}}{(\alpha/\beta)_{str}}\right]}\right\} \qquad (9.6)$$

where all the terms on the right-hand side, $(\alpha/\beta)_{str}$, d_{ref}, n_{ref} and n_{new}, are known.

In Figure 9.5(a), for $\alpha/\beta = 10$ Gy, the generic value for tumor clonogens, we see that 3 fractions of 11.1 Gy and 20 fractions of 2.75 Gy are exactly *iso*-effective, as both result in the same *SF* ($= \sim 10^{-9}$). In Figure 9.5(b), where α/β is taken as 3 Gy,

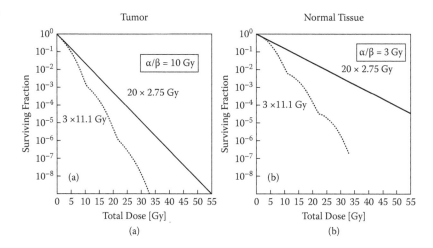

Figure 9.5. (a) Surviving fraction, *SF* (log scale), versus total dose (Gy) for conventional fractionation (20 × 2.75 Gy), full line, and for a highly hypofractionated regimen (3 × 11.1 Gy), dotted curve: with $\alpha/\beta = 10$ Gy (generic tumor) demonstrating exact *iso*-effect; (b) with $\alpha/\beta = 3$ Gy (generic normal tissue) showing much lower *SF* for the three-fraction regimen than for 20 fractions. Parameters used to generate curves: (a) $\alpha = 0.3$ Gy⁻¹, $\beta = 0.03$ Gy⁻²; (b) $\alpha = 0.1$ Gy⁻¹, $\beta = 0.033$ Gy⁻² (A. Hoffmann, private communication).

the generic value for late normal-tissue effects, it is observed that 3 fractions of 11.1 Gy and 20 fractions of 2.75 Gy are definitely *not iso*-effective. The curves indicate that the 3 × 11.1 Gy hypofractionated regimen would be significantly more damaging to normal tissue than 20 fractions of 2.75 Gy. In fact, the three-fraction regimen for *normal-tissue iso*-effectivity has a fraction size of 8.88 Gy. As this is significantly lower than 11.1 Gy, it would result in a much lower probability of tumor control. This example is perfectly consistent with Figure 9.4 (i.e., smaller fraction sizes yield a higher therapeutic ratio).

The use of Equation (9.5) or (9.6) to derive a new *iso*-effective fractionation scheme is generally unproblematic when the structure of interest is the tumor; the new prescription would be unambiguously given by D_{new} and d_{new} as in the preceding example. However, when changing fraction size/fraction number, the condition more commonly desired is *normal-tissue iso*-effect. Nevertheless, just as in the preceding example, the conventional practice is to plug the *tumor* dose prescription into the preceding formulae, despite the fact that the new prescription is to be *iso*-effective for normal tissue. Hoffmann and Nahum (2013) pointed out that this application of the Withers formula is only valid when *either* the normal tissue is irradiated uniformly with the same dose as the tumor (almost never the case in clinical practice) *or* the response of the normal tissue is determined by the fractional volume receiving the highest dose (which, for the principal organ at risk, will almost certainly be very close to that in the tumor). This latter condition corresponds to 100% "serial" behavior—for example, spinal cord (see following).

Hoffmann and Nahum (2013) have proposed a modified parameter, $(\alpha/\beta)_{eff}$, to be used in Equations (9.4)–(9.6), that reflects both the heterogeneity of the dose distribution received by the normal tissue and the degree to which it behaves in a "serial" or "parallel" manner. For "parallel" organs, such as lung with radiation pneumonitis as the end-point, and for the highly conformal treatment plans achievable today, especially for small lung tumors, $(\alpha/\beta)_{eff}$ has a much higher value than the *intrinsic* one of ≈ 3, and this $(\alpha/\beta)_{eff}$ can even approach or exceed that of the generic tumor value (i.e., 10). The extremely hypofractionated regimens for early stage non-small cell lung tumors (e.g., Baumann et al. 2009), known as SBRT or SABR, are examples of highly conformal treatments with a "parallel" organ at risk (the uninvolved, paired lung). SABR protocols involve total doses only marginally lower than those used in hyperfractionated treatments (i.e., ≈ 20–30 fractions) delivered in 3–5 fractions with acceptable toxicity; this is consistent with a high $(\alpha/\beta)_{eff}$.

What is the effective fraction size in a brachytherapy treatment? This all depends on the dose-rate. There are two main types of brachytherapy: low dose-rate (LDR) and high dose-rate (HDR). An example of the former is the insertion of radioactive seeds in the tumor (e.g., prostate); the dose is delivered over a very long period as the radioactive source decays. To a good approximation all the sublethal damage is repaired and hence β can be set to zero in the LQ expression; the *BED* is therefore equal to the total dose, integrated over time. Referring to the text which described Equation (9.2), this is equivalent to delivering the dose in a very large number of very small fractions, which should theoretically result in a high therapeutic ratio. However, during the very long overall time, considerable clonogen repopulation may occur. This is unlikely to be a problem in the case of prostate tumors but would be inadvisable for tumors in the lung or head and neck region (cf. Saunders et al. 1999). By contrast, brachytherapy administered at a high dose-rate (e.g., in pulsed mode) is radiobiologically much closer to fractionated external-beam therapy. For a detailed treatment of the radiobiology of the different modes of brachytherapy, the reader is referred to Dale (2007).

9.4   Functional Subunits (FSUs) and the Volume Effect

In contrast to the tumor situation, the response of normal tissues to irradiation cannot be simply explained by the number of cells "killed." Consider a normal tissue such as the spinal cord. A high dose to a very small part of the cord, even with negligible dose to the remainder, can result in paralysis. For the same irradiation pattern in a tumor there would be absolutely zero probability of control. The *spatial* distribution of dose to a normal tissue is crucial. This is often expressed in terms of tissue "architecture" or, put another way, the severity of radiation-induced normal-tissue damage depends on the *volume* of tissue irradiated, summarized in the term "volume effect."

A powerful idea in understanding volume effects is that of the *functional subunit*, or FSU. This is the tissue volume that can be regenerated by a single stem cell. Withers, Taylor and Maciejewski (1988) postulated that "organ function depends upon the aggregation of cells into function subunits." A damaged FSU cannot therefore be "rescued" by migration of stem cells from nearby FSUs. Thus, if all these units retain at least one undamaged stem cell, the irradiation would not cause any damage.

Figure 9.6 is a highly schematic illustration of alternative arrangements of FSUs. An arrangement like the links in a chain is known as "serial" or "in series" and is illustrated in Figure 9.6(a). Here, damage to any one FSU results in loss of organ function. The spinal cord is the clearest example of this.

In complete contrast, for a "parallel" organization, as in Figure 9.6(b), damage to a single or even to several FSUs may have no effect on organ function, as there are alternative pathways. The analogy with electrical circuits may be conceptually helpful. The lung is a clear example of a "parallel" organ; it is well known that a patient can still breathe when part or even the whole of one lung is severely damaged or even missing. Parallel organs are said to possess a large *functional* reserve.

Since a single stem cell can regenerate a whole FSU, the magnitude of the dose required to "destroy" an FSU must depend on the number of stem cells in the FSU as well as on the intrinsic radiosensitivity of these cells. Rutkowska, Baker and Nahum (2010) have developed a three-dimensional (3D) normal-tissue damage model by combining linear-quadratic radiobiology with these ideas; Rutkowska et al. (2012) applied this model specifically to radiation pneumonitis in the lung.

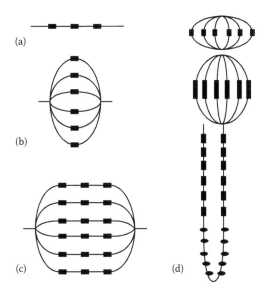

Figure 9.6. An illustration of the concepts of (a) series (or serial), for example, spinal cord; (b) parallel, for example, lungs, liver; (c) series-parallel, for example, the heart; (d) a combination of parallel and serial structures, for example, a nephron (from ICRU Report 62, 1999, modified from Withers, H. R. et al., 1988, *International Journal of Radiation Oncology Biology Physics* 14:751–759, and Källman, P., Ågren, A. and Brahme, A., 1992, *International Journal of Radiation Biology* 62:249–262.)

The relationship between the probability of damage and the distribution of dose in any normal tissue can be at least partially understood in terms of these simple ideas. Thus, in the case of the spinal cord it is the *maximum* dose that is the key quantity, and "tolerance" for this organ is more or less guaranteed by making sure that no part of the cord receives a dose greater than this *tolerance dose* (e.g., Kirkpatrick, van der Kogel and Schultheiss 2010). For lung, the situation is very different. Due to its "parallel" nature, a high dose to a small volume can be tolerated, but a relatively low dose to a large volume may not be. It has been found that the key parameter in this case, for the end-point of radiation pneumonitis, is the *mean dose* to the paired non-involved lung; that is, for a tumor situated in the lung, one subtracts the gross tumor volume (GTV) (Marks, Bentzen, et al. 2010).

It is appropriate at this point to say a few words about modern representations of the dose distributions in critical structures (target volumes such as GTV, PTV and normal tissues). The 3D distributions generated by today's radiotherapy treatment planning systems are firstly impossible to visualize except as a series of "slices" through the patient, and secondly contain a vast amount of data. This data-overload problem has been solved pragmatically by reducing the 3D dose distributions to two-dimensional (2D) representations showing how much of the volume of the structure in question receives what (total) dose—that is, a frequency distribution, known as a differential dose-volume histogram (DVH). However, it is the cumulative form of the DVH that has become the universal method of visually assessing treatment plans (e.g., Bidmead and Rosenwald 2007). Figure 9.7 shows the relationship between these two representations of a DVH.

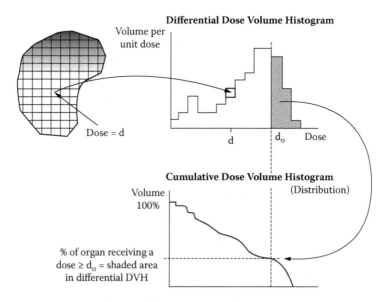

Figure 9.7. An illustration of how a voxelized dose distribution can be converted into a dose-volume histogram (DVH), shown here in its fundamental *differential* and more often used *cumulative* forms. (Redrawn from Goitein, M. 1992. *Seminars in Radiation Oncology* 2:246–256.)

Any point on the (cumulative) DVH yields the volume (often shown as a percentage of the total volume of the structure) receiving *at least that dose*. Normal-tissue constraints are frequently specified in terms of $V_{XXGy} \leq YY\%$, which translates into *the volume receiving XX Gy or greater should not exceed YY% of the total volume* (see Table 9.1 at the end of this chapter).

Returning to the subject of volume effects, tissue tolerance is frequently assumed to be described by a power law function of the fractional volume irradiated (e.g., Marks, Yorke, et al. 2010):

$$D(V_{irradiated}) = D(V_{reference}) \left(\frac{V_{reference}}{V_{irradiated}} \right)^n \qquad (9.7)$$

where $V_{reference}$ is the reference volume and $V_{irradiated}$ is the uniformly irradiated (normal-tissue) volume, usually expressed as partial (i.e., fractional) volumes. The $D(V_{reference})$ is frequently the corresponding tolerance dose, representing a chosen level on the dose-response curve, such as TD_{50} (the dose level resulting in a 50% probability of damage). The parameter n controls the volume effect. Referring to the examples discussed above, $n \approx 1$ for a parallel organ such as lung, and $n \approx 0.1$ for a series organ such as the spinal cord.

Lyman (1985) used this power law model to define the risks associated with uniform partial-volume irradiation. From Equation (9.7), *decreasing* the irradiated volume fraction shifts the dose-response curve (TD_{50}) to *higher* doses by the ratio of the irradiated fractional volumes, ($V_{irradiated}/V_{reference}$), raised to the power $-n$. For example, if n equals 1, then, relative to the TD_{50} for whole-volume irradiation, TD_{50} (1), the TD_{50} for one-half of the volume, TD_{50} (0.5), is expected to increase by a factor of 2, whereas if $n = 0.5$, TD_{50} (0.5) would increase only by a factor $\sqrt{2}$.

To generalize this to clinically realistic (i.e., *heterogeneous*) dose distributions, the generalized equivalent uniform dose (*gEUD*) is commonly employed (Niemierko 1999). The *gEUD* is defined as the dose given *uniformly* to the entire normal tissue or organ yielding the *same complication rate* as the (partial, heterogeneous) dose distribution the structure in question actually receives. The *gEUD* is computed by summing over all the voxels i in the structure:

$$gEUD(d_i, V_i, n) = \left[\frac{1}{N_{voxels}} \left\{ d_1^{1/n} + d_2^{1/n} + d_3^{1/n} + \ldots + d_{N_{voxels}}^{1/n} \right\} \right] \qquad (9.8)$$

where N_{voxels} is the number of voxels (of equal volume), and d_i is the dose to the *i*th voxel.

It should be noted that the *gEUD* equation is consistent with the power-law assumption, with the same parameter n appearing in Equations (9.7) and (9.8). Together, the *gEUD* equation and the Lyman assumptions are virtually equivalent to the Lyman–Kutcher–Burman (LKB) NTCP model (Lyman 1985; Kutcher et al. 1991); sometimes the parameter a, equal to $1/n$, is employed. When n is small (a is large), changes in irradiated volume result in only modest changes in

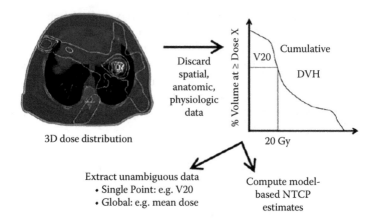

Figure 9.8. A three-dimensional dose distribution is reduced to a two-dimensional (2D) dose-volume histogram (DVH) by discarding all spatial, anatomic and physiologic data. The 2D graph is then further reduced to a single value of merit, such as the mean dose, the percent of the organ receiving ≥20 Gy (V_{20}), or a model-based normal-tissue complication probability (NTCP). (Reproduced with permission from Marks, L. B., Ten Haken, R. T., et al. 2010. *International Journal of Radiation Oncology, Biology, Physics* 76 (no. 3, supplement).)

relative tolerance (e.g., spinal cord), whereas, for large n (small a), the tolerance dose depends strongly on the irradiated volume fraction; lung is an excellent example of the latter, being said to exhibit a large volume effect.

Figure 9.8 can serve as a summary of much of the preceding material. The complex 3D dose distribution in a normal tissue (the lung is suggested in the figure) is reduced to a (cumulative) DVH, from which a single-point metric such as V_{20} can be picked off. This metric may be used to assess the tolerance of the treatment plan; alternatively, a model-based NTCP may be calculated (e.g., Kutcher et al. 1991; Nahum and Kutcher 2007).

9.5   A Summary of the QUANTEC Study

As is apparent from the previous sections, while radiobiological principles can act as a guide, with fractionation as the clearest example, quantification of normal-tissue damage following radiation therapy is essentially a matter of empirical observation. Three-dimensional radiotherapy treatment planning, which arrived around the mid-1980s, has furnished us with detailed information not only about doses but also about *distributions* of dose in normal tissues (usually in the form of DVHs). Linking this step-change in our knowledge of the "input" to the radiotherapy process (doses, volumes, numbers of fractions, etc.), with the "output" in terms of patient follow-up (normal-tissue complication rates, as well as tumor control and cure rates) will yield a treasure trove of knowledge.

A major (several-year long) collaborative effort to pool this knowledge, badged as *quantitative analyses of normal-tissue effects in the clinic* (QUANTEC), resulted in a dedicated issue of the *Internal Journal of Radiation Oncology, Biology, Physics* (Marks, Ten Haken, et al. 2010) including papers on all the major *organs at risk,* including brain, spinal cord, the parotid glands, lung, heart, esophagus, liver, kidney, bladder and rectum. In Table 9.1 we reproduce the data in the table headed "QUANTEC Summary: Approximate Dose/Volume/Outcome Data for Several Organs following Conventional Fractionation" from the paper "Use of Normal Tissue Complication Probability Models in the Clinic" (Marks, Yorke, et al. 2010).

Table 9.1 contains guidelines for "safe" radiotherapy (i.e., respecting the specified dose-volume constraints should ensure acceptably low complication rates). However, no figures are given for the expected complication rates at the threshold value (i.e., an NTCP value) and no information is given on the expected *rate of increase* in complication frequency if one exceeds the constraint. Such information is the job of full-blown NTCP models, which will be discussed in detail in the volume planned to follow on from this one.

The caveat *following conventional fractionation* cannot be overemphasized. Conventional fractionation implies small fractions of 1.8–2.0 Gy delivered to the target volume. Consider $V_{20} \leq 30\%$ for the lung; this translates into "the relative volume (of the non-involved lung) at the 20 Gy point in the cumulative dose-volume histogram (cf. Figure 9.8) should not exceed 30%." However, in order to assess the radiobiological effect of a specific dose, here 20 Gy or greater, one needs to know the *number of fractions* of the treatment. This key piece of information is not apparent from "following conventional fractionation," which only tells us the *size of the dose per fraction delivered to the tumor.* Therefore constraints such as $V_{XX} \leq YY\%$ ought to refer to a DVH that has been corrected to a specific fraction size, such as 2.0 Gy, as is done in the Lyman–Kutcher–Burman NTCP model (e.g., Nahum and Kutcher 2007).

Table 9.1. QUANTEC Summary: Approximate Dose/Volume/Outcome Data for Several Organs Following Conventional Fractionation (Unless Otherwise Noted)*

Organ	Volume segmented	Irradiation type (partial organ unless otherwise stated)[†]	Endpoint	Dose (Gy), or dose/volume parameters[†]	Rate (%)	Notes on dose/ volume parameters
Brain	Whole organ	3D-CRT	Symptomatic necrosis	Dmax <60	<3	Data at 72 and 90 Gy, extrapolated 5 from BED models
	Whole organ	3D-CRT	Symptomatic necrosis	Dmax = 72	5	
	Whole organ	3D-CRT	Symptomatic necrosis	Dmax = 90	10	
	Whole organ	SRS (single fraction)	Symptomatic necrosis	V12 <5–10 cc	<20	Rapid rise when V12 > 5–10 cc
Brain stem	Whole organ	Whole organ	Permanent cranial neuropathy or necrosis	Dmax <54	<5	
	Whole organ	3D-CRT	Permanent cranial neuropathy or necrosis	D1–10 cc[‖] ≤59	<5	
	Whole organ	3D-CRT	Permanent cranial neuropathy or necrosis	Dmax <64	<5	Point dose <<1 cc
	Whole organ	SRS (single fraction)	Permanent cranial neuropathy or necrosis	Dmax <12.5	<5	For patients with acoustic tumors
Optic nerve/ chiasm	Whole organ	3D-CRT	Optic neuropathy	Dmax <55	<3	Given the small size, 3D CRT is often whole organ[‡]
	Whole organ	3D-CRT	Optic neuropathy	Dmax 55–60	3–7	
	Whole organ	3D-CRT	Optic neuropathy	Dmax >60	>7–20	
	Whole organ	SRS (single fraction)	Optic neuropathy	Dmax <12	<10	
Spinal cord	Partial organ	3D-CRT	Myelopathy	Dmax = 50	0.2	Including full cord cross-section
	Partial organ	3D-CRT	Myelopathy	Dmax = 60	6	
	Partial organ	3D-CRT	Myelopathy	Dmax = 69	50	

(Continued)

Table 9.1. *(Continued)* QUANTEC Summary: Approximate Dose/Volume/Outcome Data for Several Organs Following Conventional Fractionation (Unless Otherwise Noted)*

Organ	Volume segmented	Irradiation type (partial organ unless otherwise stated)†	Endpoint	Dose (Gy), or dose/volume parameters†	Rate (%)	Notes on dose/volume parameters
Spinal cord	Partial organ	SRS (single fraction)	Myelopathy	Dmax = 13	1	Partial cord cross-section irradiated 3 fractions,
	Partial organ	SRS (hypofraction)	Myelopathy	Dmax = 20	1	partial cord cross-section irradiated
Cochlea	Whole organ	3D-CRT	Sensory neural hearing loss	Mean dose ≤45	<30	Mean dose to cochlear, hearing at 4 kHz
	Whole organ	SRS (single fraction)	Sensory neural hearing loss	Prescription dose ≤14	<25	Serviceable hearing
Parotid	Bilateral whole parotid glands	3D-CRT	Long term parotid salivary function reduced to <25% of pre-RT level	Mean dose <25	<20	For combined parotid glands¶
	Unilateral whole parotid gland	3D-CRT	Long term parotid salivary function reduced to <25% of pre-RT level	Mean dose <20	<20	For single parotid gland. At least one parotid gland spared to <20 Gy¶
	Bilateral whole parotid glands	3D-CRT	Long term parotid salivary function reduced to <25% of pre-RT level	Mean dose <39	<50	For combined parotid glands (per Fig. 3 in Marks, Yorke et al. 2010)¶
Pharynx	Pharyngeal constrictors	Whole organ	Symptomatic dysphagia and aspiration	Mean dose <50	<20	Based on Section B4 in Marks, Yorke et al. 2010

(Continued)

Table 9.1. (Continued) QUANTEC Summary: Approximate Dose/Volume/Outcome Data for Several Organs Following Conventional Fractionation (Unless Otherwise Noted)[*]

Organ	Volume segmented	Irradiation type (partial organ unless otherwise stated)[†]	Endpoint	Dose (Gy), or dose/volume parameters[†]	Rate (%)	Notes on dose/ volume parameters
Larynx	Whole organ	3D-CRT	Vocal dysfunction	Dmax <66	<20	With chemotherapy, based on single study (see Section A4.2 in Marks, Yorke et al. 2010)
	Whole organ	3D-CRT	Aspiration	Mean dose <50	<30	With chemotherapy, based on single study (see Fig. 1 in Marks, Yorke et al. 2010)
	Whole organ	3D-CRT	Edema	Mean dose <44	<20	Without chemotherapy,
	Whole organ	3D-CRT	Edema	V50 <27%	<20	based on single study in patients without larynx cancer[**]
Lung	Whole organ	3D-CRT	Symptomatic pneumonitis	V20 ≤ 30%	<20	For combined lung. Gradual dose response
	Whole organ	3D-CRT	Symptomatic pneumonitis	Mean dose = 7	5	Excludes purposeful whole lung irradiation
	Whole organ	3D-CRT	Symptomatic pneumonitis	Mean dose = 13	10	
	Whole organ	3D-CRT	Symptomatic pneumonitis	Mean dose = 20	20	
	Whole organ	3D-CRT	Symptomatic pneumonitis	Mean dose = 24	30	
	Whole organ	3D-CRT	Symptomatic pneumonitis	Mean dose = 27	40	
Esophagus	Whole organ	3D-CRT	Grade ≥3 acute esophagitis	Mean dose <34	5–20	Based on RTOG and several studies

(Continued)

Table 9.1. (*Continued*) QUANTEC Summary: Approximate Dose/Volume/Outcome Data for Several Organs Following Conventional Fractionation (Unless Otherwise Noted)*

Organ	Volume segmented	Irradiation type (partial organ unless otherwise stated)†	Endpoint	Dose (Gy), or dose/volume parameters†	Rate (%)	Notes on dose/volume parameters
Esophagus	Whole organ	3D-CRT	Grade ≥2 acute esophagitis	V35 <50%	<30	A variety of alternate threshold doses have been implicated.
	Whole organ	3D-CRT	Grade ≥2 acute esophagitis	V50 <40%	<30	Appears to be a dose/volume response
	Whole organ	3D-CRT	Grade ≥2 acute esophagitis	V70 <20%	<30	
Heart	Pericardium	3D-CRT	Pericarditis	Mean dose <26	<15	Based on single study
	Pericardium	3D-CRT	Pericarditis	V30 <46%	<15	
	Whole organ	3D-CRT	Long-term cardiac mortality	V25 <10%	<1	Overly safe risk estimate based on model predictions
Liver	Whole liver – GTV	3D-CRT or Whole organ	Classic RILD††	Mean dose <30–32	<5	Excluding patients with pre-existing liver disease or hepatocellular carcinoma, as tolerance doses are lower in these patients
	Whole liver – GTV	3D-CRT	Classic RILD	Mean dose <42	<50	

(*Continued*)

Table 9.1. (Continued) QUANTEC Summary: Approximate Dose/Volume/Outcome Data for Several Organs Following Conventional Fractionation (Unless Otherwise Noted)*

Organ	Volume segmented	Irradiation type (partial organ unless otherwise stated)†	Endpoint	Dose (Gy), or dose/volume parameters†	Rate (%)	Notes on dose/volume parameters
Liver	Whole liver – GTV	3D-CRT or Whole organ	Classic RILD	Mean dose <28	<5	In patients with Child-Pugh A preexisting liver disease or hepatocellular carcinoma, excluding hepatitis B reactivation as an endpoint
	Whole liver – GTV	3D-CRT	Classic RILD	Mean dose <36	<50	
	Whole liver – GTV	SBRT (hypofraction)	Classic RILD	Mean dose <13	<5	3 fractions, for primary liver cancer
	Whole liver – GTV	SBRT (hypofraction)	Classic RILD	Mean dose <18	<5	6 fractions, for primary liver cancer
	Whole liver – GTV	SBRT (hypofraction)	Classic RILD	Mean dose <15	<5	3 fractions, for liver metastases
	Whole liver – GTV	SBRT (hypofraction)	Classic RILD	Mean dose <20	<5	6 fractions, for liver metastases
	>700 cc of normal liver	SBRT (hypofraction)	Classic RILD	Dmax <15	<5	Critical volume based, in 3–5 fractions

(Continued)

Table 9.1. (*Continued*) QUANTEC Summary: Approximate Dose/Volume/Outcome Data for Several Organs Following Conventional Fractionation (Unless Otherwise Noted)*

Organ	Volume segmented	Irradiation type (partial organ unless otherwise stated)[†]	Endpoint	Dose (Gy), or dose/volume parameters[†]	Rate (%)	Notes on dose/volume parameters
Kidney	Bilateral whole kidney[‡]	Bilateral whole organ or 3D-CRT	Clinically relevant renal dysfunction	Mean dose <15–18	<5	
	Bilateral whole kidney[‡]	Bilateral whole organ	Clinically relevant renal dysfunction	Mean dose <28	<50	
	Bilateral whole kidney[‡]	3D-CRT	Clinically relevant renal dysfunction	V12 <55%, V20 <32%, V23 <30%, V28 <20%	<5	For combined kidney
Stomach	Whole organ	Whole organ	Ulceration	D100[‖] <45	<7	
Small bowel	Individual small bowel loops	3D-CRT	Grade ≥3 acute toxicity[§]	V15 <120 cc	<10	Volume based on segmentation of the individual loops of bowel, not the entire potential peritoneal space
	Entire potential space within peritoneal cavity	3D-CRT	Grade ≥3 acute toxicity[§]	V45 <195 cc	<10	Volume based on the entire potential space within the peritoneal cavity

(*Continued*)

Table 9.1. *(Continued)* QUANTEC Summary: Approximate Dose/Volume/Outcome Data for Several Organs Following Conventional Fractionation (Unless Otherwise Noted)*

Organ	Volume segmented	Irradiation type (partial organ unless otherwise stated)†	Endpoint	Dose (Gy), or dose/volume parameters†	Rate (%)	Notes on dose/ volume parameters
Rectum	Whole organ	3D-CRT	Grade ≥2 late rectal toxicity,	V50 <50%	<15	Prostate cancer treatment
			Grade ≥3 late rectal toxicity		<10	
	Whole organ	3D-CRT	Grade ≥2 late rectal toxicity,	V60 <35%	<15	
			Grade ≥3 late rectal toxicity		<10	
	Whole organ	3D-CRT	Grade ≥2 late rectal toxicity,	V65 <25%	<15	
			Grade ≥3 late rectal toxicity		<10	
	Whole organ	3D-CRT	Grade ≥2 late rectal toxicity,	V70 <20%	<15	
			Grade ≥3 late rectal toxicity		<10	
	Whole organ	3D-CRT	Grade ≥2 late rectal toxicity,	V75 <15%	<15	
			Grade ≥3 late rectal toxicity		<10	
Bladder	Whole organ	3D-CRT	Grade ≥3 late RTOG	Dmax <65	<6	Bladder cancer treatment. Variations in bladder size/ shape/ location during RT hamper ability to generate accurate data
	Whole organ	3D-CRT	Grade ≥3 late RTOG	V65 ≤50 %		Prostate cancer treatment
				V70 ≤35 %		Based on current RTOG
				V75 ≤25 %		0415 recommendation
				V80 ≤15 %		

(Continued)

Table 9.1. (Continued) QUANTEC Summary: Approximate Dose/Volume/Outcome Data for Several Organs Following Conventional Fractionation (Unless Otherwise Noted)*

Organ	Volume segmented	Irradiation type (partial organ unless otherwise stated)†	Endpoint	Dose (Gy), or dose/volume parameters†	Rate (%)	Notes on dose/volume parameters
Penile bulb	Whole organ	3D-CRT	Severe erectile dysfunction	Mean dose to 95% of gland <50	<35	
	Whole organ	3D-CRT	Severe erectile dysfunction	D90‖ <50	<35	
	Whole organ	3D-CRT	Severe erectile dysfunction	D60–70 <70	<55	

Source: (Reproduced with permission from Marks, L. B., Yorke, E. D., et al. 2010. *International Journal of Radiation Oncology, Biology, Physics* 76 (no. 3, supplement) S10–S19)

Abbreviations: 3D-CRT = 3-dimensional conformal radiotherapy, SRS = stereotactic radiosurgery, BED = Biologically effective dose, SBRT = stereotactic body radio-therapy, RILD = radiation-induced liver disease, RTOG = Radiation Therapy Oncology Group.

* All data are estimated from the literature summarized in the QUANTEC reviews unless otherwise noted. Clinically, these data should be applied with caution. Clinicians are strongly advised to use the individual QUANTEC articles to check the applicability of these limits to the clinical situation at hand. They largely do not reflect modern IMRT.

† All at standard fractionation (i.e., 1.8–2.0 Gy per daily fraction) unless otherwise noted. Vx is the volume of the organ receiving ≥ x Gy. Dmax = Maximum radiation dose.

‡ Non-TBI.

§ With combined chemotherapy.

‖ Dx = minimum dose received by the "hottest" x% (or x cc's) of the organ.

¶ Severe xerostomia is related to additional factors including the doses to the submandibular glands.

** Estimated by Dr. Eisbruch.

†† Classic Radiation induced liver disease (RILD) involves anicteric hepatomegaly and ascites, typically occurring between 2 weeks and 3 months after therapy. Classic RILD also involves elevated alkaline phosphatase (more than twice the upper limit of normal or baseline value).

‡‡ For optic nerve, the cases of neuropathy in the 55 to 60 Gy range received ≈59 Gy (see Mayo, Martel et al. 2010 for details). Excludes patients with pituitary tumors where the tolerance may be reduced.

10 Radiobiology Applied to Tumor Response Modeling

The description of frontline tumor control probability (TCP) and normal tissue complication probability (NTCP) modeling will be the subject of Volume II of this textbook. As a bridge between the basic radiobiology of Volume I and the "macroscopic" modeling of clinical outcomes in Volume II, this chapter will describe why the basic biological parameters are important for modeling radiotherapy response. Local control is thought to occur when all the stem cells in the tumor are killed by the radiation treatment. And since the microdosimetric distributions of inactivating energy depositions are random (stochastic), Poisson statistics are again employed to calculate the proportion of the various tumor types in which every stem cell is inactivated (i.e., the TCP) as a function of dose. For these TCP values to predict clinical response accurately, the values of intrinsic radiosensitivity of the stem cells and their total number in the tumor (i.e., the cells that must be inactivated) should be known. These parameters are not usually determined for individual tumors but reasonable estimates are available for populations of like tumors in different patients. Intertumor heterogeneity should be taken into account when modeling the radiation response of groups of human cancers. This heterogeneity arises from distributions in clonogen intrinsic radiosensitivity, oxygenation status, growth fraction and possibly other factors. In addition, host factors such as immune response can also play a role. TCP models are important for determining the effectiveness of novel fractionation schemes, the effect of departures from homogeneous irradiation, variations in initial clonogen number and other factors.

10.1　Surviving Fractions after Fractionated Dose Delivery

When radiation is administered in daily fractions of about 1.8–3.0 Gy, tumor cell survival will be described by Equation (5.1) in Chapter 5, provided that there is complete repair of sublethal lesions during the interfraction intervals (which should not be less than 6 hr) and that no significant proliferation occurs. Figure 10.1 shows such survival curves after daily 2 Gy dose fractions for aerobic tumor cells whose average $\bar{\alpha}$-inactivation values are 0.73, 0.36 and 0.26 Gy^{-1} and whose $\bar{\beta}_0$-inactivation value is constant at 0.0581 Gy^{-2} (see Table 7.2, Chapter 7) (Chapman 2014). Also shown are the survival curves for the same tumor cells if they are hypoxic throughout the fractionated treatment (OERs of 1.8 and 3.0 were assumed for $\bar{\alpha}$- and $\sqrt{\bar{\beta}_0}$-parameters, respectively). It is apparent that total doses of 30–65 Gy will kill up to 10 logs of aerobic cells (i.e., 10^{10} cells) in tumors of groups A and B, groups C and D, and group E, and even up to 10 logs of hypoxic cells of groups A and B. Total doses of over 100 Gy are required to inactivate cells of groups C and D and group E if they are hypoxic throughout the treatment. These input parameters (doses, radiosensitivities, initial clonogen numbers) are consistent with the known outcomes of radiation therapy. Since several animal tumors exhibit initial hypoxic fractions (HFs) of 5%–50%, there could be several logs of these radioresistant cells in human tumors if they have similar HFs. Tumor reoxygenation could then be a dominant factor for determining the probability of tumor cure and this view has dominated research over the past 40 years (Chapman 1997).

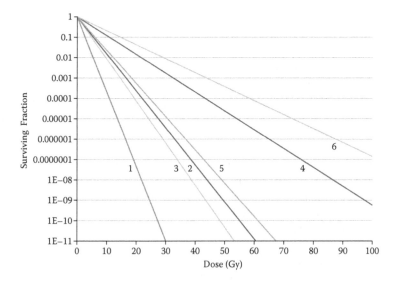

Figure 10.1. The surviving fraction of aerobic and hypoxic tumor clonogens irradiated with 2 Gy fractions whose $\bar{\alpha}$- and $\sqrt{\bar{\beta}}$-inactivation parameters are those of groups A and B (1-aerobic, 2-hypoxic), groups C and D (3-aerobic, 4-hypoxic) and group E (5-aerobic, 6-hypoxic) tumors of Table 7.2 in Chapter 7.

Animal tumor studies of tumor reoxygenation after various radiation treatments have yielded widely variable results (Hall 2000). Reoxygenation of human tumors during and after radiotherapy is poorly understood and awaits a reliable assay for measuring tumor pO_2 (Chapman, Schneider, et al. 2001).

10.2   An LQ-Based TCP Model

For TCP modeling, the number of clonogens in the tumors of interest is an essential input parameter (Buffa, Fenwick and Nahum 2000; Yaromina et al. 2007). Just what that number is for different tumor types and how it varies between tumors in different patients is not explicitly known. We do know that the gross tumor volumes (GTVs) of solid tumors in different patients can vary widely. The studies of West et al. (1993) and Björk-Eriksson et al. (2000) showed that only 0.05%–0.5% of all viable cells released by enzymes from cervical and head and neck tumors, respectively, formed colonies in soft agar assays (Courtney and Mills 1978). If these are the cells that require inactivation for tumor control and cure, they constitute a small fraction of the tumor mass. In studies directed to cancer drug development, Baker and Sanger (1991) measured the solid tumor stem cell density to be ~0.5×10^6/g (or ~0.5×10^6 cm^{-3} as the density is very close to 1 g cm^{-3}). Since radiotherapy treats tumor volumes between 0.01 and 1.0 L, the clonogen numbers (N_o) that require inactivation will be of the order of 10^7–10^9. Consequently, the previous discussion of tumor cures resulting from 10 logs of cell killing is not unreasonable. However, Figure 10.5 will demonstrate that most human tumor (population) responses to radiation can be fitted by a TCP model with a broad range of assumed cell-inactivation and initial cell-number parameter sets (from a low $\bar{\alpha}$ and an absurdly small N_o to $\bar{\alpha}$ values like those in Table 7.2 (Chapter 7) and N_o values of 10^7–10^9). This is one reason that the authors' approach has been to narrow down the range of input parameters so that best estimates based upon tumor biology measurements are used.

The LQ model can be used to generate TCP curves of various tumors starting with the intrinsic clonogen radiosensitivity and the initial number of their clonogens. Thus, we have

$$N_s = N_o \exp\left\{-\bar{\alpha}D\left(1+\left[\bar{\beta}_o / \bar{\alpha}\right]d\right)\right\} \qquad (10.1)$$

where
  N_s and N_o are the surviving and initial number of tumor clonogens, respectively,
  $\bar{\alpha}$ and $\bar{\beta}_o$ are the clonogen radiosensitivity parameters,
  D is the total dose,
  d is the fraction size.

Equation (10.1) is a different form of Equation (5.1) in Chapter 5. The Poisson probability of there being *no surviving cells* in a population of like tumors after a fractionated treatment is given by (Nahum and Sanchez-Nieto 2001)

$$TCP = \exp\{-N_s\}. \qquad (10.2)$$

It therefore follows from Equation (10.1) that

$$TCP = \exp\left(-N_o \exp\left\{-\bar{\alpha}D\left(1+\left[\bar{\beta}_o / \bar{\alpha}\right]d\right)\right\}\right) \tag{10.3}$$

where all the parameters are as previously defined.

10.3   Population Averaging

When the aerobic and hypoxic radiosensitivity parameters utilized in generating Figure 10.1 are input into Equation (10.3) along with an initial clonogen number of 10^8, the curves shown in Figure 10.2 are obtained. These data again indicate that tumor cures should be expected with total doses of 60 Gy (2–3 Gy fraction size) for aerobic and hypoxic tumors in groups A and B and aerobic tumors of groups C and D. Hypoxic tumors of groups C and D and group E will require doses of over 90 Gy to achieve local control.

These TCP curves have been generated with single values of intrinsic radiosensitivity, a situation that cannot possibly exist when different tumors in different people are pooled to yield treatment outcomes (Bentzen et al. 1990; Bentzen, Thames and Overgaard 1992; Zaider and Hanin 2011). The positions of these response curves along the dose axis will also depend upon the initial number of clonogens (N_o) that require inactivation. Figure 10.3 shows TCP curves for aerobic tumors of groups C and D for initial clonogen numbers of 10^4, 10^6, 10^8

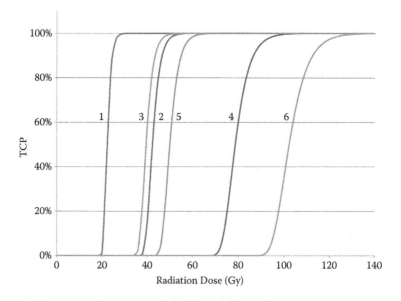

Figure 10.2. The tumor control probability (TCP) for aerobic (lines 1, 3, 5) and hypoxic (lines 2, 4, 6) tumors with radiosensitivities of groups A and B, groups C and D and group E tumors respectively (of Table 7.2, Chapter 7) for 10^8 clonogens, calculated from Equation (10.3).

(curve 3 in Figure 10.2), 10^{10} and 10^{12}. As the number of initial clonogens increases, the TCP curves will necessarily shift to higher doses. While reasonable values of intrinsic radiosensitivity for most tumors can be obtained from cell-line and tumor-biopsy studies (see Table 7.2, Chapter 7), the number of clonogens in various cancer pathologies is less well known.

For clinical tumor-response data, a much more heterogeneous situation should be expected. TCP curves derived from populations of human tumors that have been similarly treated include different GTVs, different initial clonogen numbers, different intrinsic radiosensitivites, different oxygenation status and different growth fractions. Furthermore, the variation in clinical prescriptions (over time and from clinic to clinic, country to country) rarely covers a wide range of radiation doses, so complete TCP curves are never realized. What is apparent, however, is that the slopes of TCP versus dose curves derived from clinical data are much less steep than those of tumors with homogeneous parameters i.e., those shown in Figures 10.2 and 10.3.

Nahum and colleagues (Nahum and Tait 1992; Webb and Nahum 1993; Sanchez-Nieto and Nahum 1999; Nahum and Sanchez-Nieto 2001) and Daşu, Toma-Daşu and Fowler 2003 have accounted for the intertumor variation of intrinsic radiosensitivity in a patient population by incorporating a Gaussian

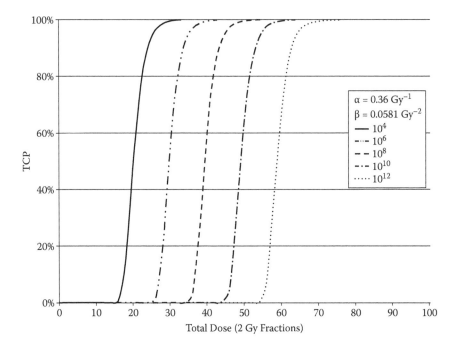

Figure 10.3. TCP vs. Dose for aerobic tumors with intrinsic radiosensitivities of groups C and D (Table 7.2, Chapter 7) whose initial clonogen numbers are 10^4, 10^6, 10^8 (curve 3 in Figure 10.2), 10^{10} and 10^{12} (from leftmost to rightmost) evaluated from Equation (10.3).

distribution function (of α-values, denoted by σ_α) into their TCP model; it is the α-killing mechanism that accounts for the majority of cell killing by standard fraction sizes. The resulting "Marsden" TCP model (see preceding references) yields a much better fit to several sets of human tumor response data, as illustrated schematically in Figure 10.4.

Since the $\bar{\beta}_o$ mechanism accounts for a much smaller proportion of the total cell killing produced by typical clinical fraction sizes and since it exhibits much less variation for different tumor cell lines than the $\bar{\alpha}$-mechanism (see Table 7.1, Chapter 7), it is less important to include a distribution function for this parameter (in practice the Marsden model keeps $\bar{\alpha}/\bar{\beta}_o$ constant as $\bar{\alpha}$ varies). While this might be acceptable for modeling tumor response to conventional fraction sizes (i.e., ≤3 Gy), it could become problematic when considering severe hypofractionation. Modeling of the *population* TCP response based on the LQ equation should include distributions of intrinsic radiosensitivity parameters (at least for single-hit killing) to account for the expected intertumor variation. This applies to all patient cohorts that describe large institutional studies as well as those by clinical cooperative groups. However, should the value of $\bar{\alpha}$ be known for an individual patient/tumor, then this value should be used directly in Equation (10.3), there being no need in such a case for a distribution of this parameter over the population (i.e., $\sigma_\alpha = 0$).

Several widely differing sets of parameters (for the "Marsden" TCP model) will adequately fit clinically observed local control versus (total) dose data.

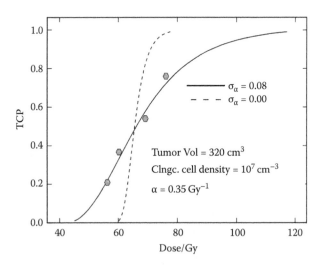

Figure 10.4. Tumor control probability (TCP) as a function of (total) dose for a representative clinical tumor with $\alpha = 0.35$ Gy^{-1} and $N_o = 3.2 \times 10^9$; the dashed curve is calculated from Equation (10.3), assuming a quasi-infinite $\bar{\alpha}/\bar{\beta}_o$ ratio. The full curve incorporates intertumor variation in $\bar{\alpha}$ through $\sigma_\alpha = 0.08$ Gy^{-1} and now fits clinical outcome data, represented by the "fictitious" data points at doses from 55 to 75 Gy. (Adapted from Webb, S. and Nahum, A. E. 1993. *Physics in Medicine and Biology* 38:653–666.)

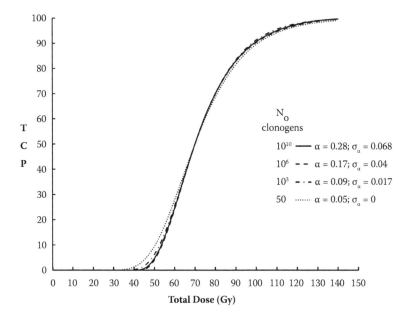

Figure 10.5. Tumor control probability (TCP) as a function of (total) dose for 4 different parameter sets corresponding to 4 different values for the initial clonogen number N_o, fitted to a fictitious clinical reference curve. a. for $N_o = 10^{10}$: $\alpha = 0.281$ Gy^{-1}, $\sigma_\alpha = 0.068$ Gy^{-1} (dashed); b. for $N_o = 10^6$: $\alpha = 0.1712$ Gy^{-1}, $\sigma_\alpha = 0.040$ Gy^{-1} (full line); c. for $N_o = 10^3$: $\alpha = 0.08805$ Gy^{-1}, $\sigma_\alpha = 0.017$ Gy^{-1} (dashed-dotted); d. for $N_o = 50$: $\alpha = 0.051$ Gy^{-1}, $\sigma_\alpha = 0$ (dotted). The fraction size is 2.0 Gy, $\alpha/\beta = 10$ Gy and the dose was assumed to be perfectly uniform.

This is illustrated in Figure 10.5. Four different assumptions have been made about the initial number of clonogens, from 10^{10} down to the unrealistically low figure of just 50. Corresponding to each of these N_o, one can find values of α and σ_α which yield TCP versus dose curves that are indistinguishable (here, $\alpha/\beta = 10$ Gy and the fraction size is 2.0 Gy). The message here is that it is possible to obtain completely meaningless values of tumor-clonogen radiosensitivity from fitting a TCP model to clinical outcome data; this should be noted by those researchers who persist in publishing values of α an order of magnitude lower than values obtained experimentally (see earlier chapters).

10.4   Incorporating Tumor Biology

Modeling prostate response to various treatments has been addressed by several groups since the brachytherapy of Stock et al. (1998) and the conformal treatments of Hanks et al. (1997) investigated a relatively wide range of doses. It is not possible to derive exact values of intrinsic radiosensitivity or initial clonogen numbers from such clinical response data. However, a realistic estimate of at least one of these parameters is essential to obtain meaningful TCP responses. In order to fit the Stock et al. dose-response curve, Brenner and Hall (1999) arrived at an α-inactivation parameter of 0.03 Gy^{-1} (β-inactivation was assumed to be negligible for low-dose brachytherapy) together with an initial clonogen number ~15. By contrast, Nahum et al. (2003) used the average value of radiosensitivity parameters from published prostate cancer cell lines together with a clonogen number of 5×10^6. This estimate for the initial clonogen number is consistent with a clonogen density of $\approx 5 \times 10^5$/g, as discussed before, and with what pathologists and urologists suggest is appropriate. Additionally, the Nahum et al. 2003 study incorporated new experimental information about the hypoxic status of prostate cancers presenting for brachytherapy at the Fox Chase Cancer Center (Movsas et al. 2000, 2002). This hypoxia information was also used to model the Hanks external-beam outcome data stratified according to pretreatment PSA level. Nahum et al. (2003) were able to make sense of both the brachy- and external-beam therapy outcome data by utilizing *in vitro* $\overline{\alpha}$- and $\overline{\beta}_0$-inactivation parameters, including their population variation over the different human prostate tumor cell lines. Prostate cancer cell lines exhibit $\overline{\alpha}$-inactivation parameters of ~0.25 Gy^{-1}, a value that places them in group E of the values shown in Table 7.2 (Chapter 7). Attempting to extrapolate $\overline{\alpha}$-inactivation parameters from clinical data will always include heterogeneities associated with tumor physiology and biology (Valdagni et al. 2005). Again, we emphasize that tumor response modeling will be most informative if it incorporates all the known factors of tumor biology that might impact treatment outcomes.

It is evident from Figure 7.3 (Chapter 7) and Figure 10.1 that the oxygenation status of clonogens in human tumors can potentially have a profound effect on tumor response. Unfortunately, no standardized method is available today to provide this information to the prescribing physicians. The Eppendorf electrode device was used by Movsas et al. (2000, 2002) to indicate that a significant proportion of men coming for brachytherapy had prostate tissue that was extremely hypoxic. This finding was confirmed using somewhat different procedures by Milosevic et al. (2012). Consequently, in the Nahum et al. (2003) TCP analysis, prostate tumors were classified as either aerobic or hypoxic at the beginning and throughout their treatments with brachytherapy or conformal (external-beam) therapy. This partitioning of tumors according to initial oxygenation status yielded TCP predictions consistent with the outcomes recorded in the extensive Hanks database (Nahum et al. 2003).

This research also suggested how important prior knowledge of tumor biology parameters that impact upon clonogen intrinsic sensitivity is for obtaining meaningful results. Solid human tumors can exhibit heterogeneity of intrinsic radiosensitivity parameters and of oxygenation status. The Chapman laboratory identified a technique for quantifying solid tumor hypoxic fractions with bioreducible chemicals that become covalently bound to molecules selectively in hypoxic cells (Chapman, Franko and Sharplin 1981). The chemical adducts could be identified by nuclear medicine (Parliament et al. 1992; Groshar et al. 1993), by fluorescent antibody sensing of marker adducts in biopsy specimens (Raleigh, Miller and Franko 1987) and by nuclear magnetic resonance techniques (Raleigh et al. 1991). The Chapman laboratory emphasized the visualization of hypoxia in solid tumors *in vivo* by nuclear medicine procedures of single-photon emission tomography (SPECT) or positron emission tomography (PET) since these are imaging platforms that are currently utilized in cancer treatment planning. To our knowledge, none of the suggested procedures for imaging tumor hypoxia based on this bioreducible chemical method is in routine use today. With such a validated technique, the oxygenation status of human tumors prior to, during and after radiotherapy could be measured to assist with devising improved treatment strategies. Urtasun et al. (1998) reported preliminary data for small lung cancers that suggested that the presence of viable hypoxic tumor at the end of cancer treatment predicted early treatment failures. Over 1,000 cancer patient tumors have now been imaged with F-MISO but there has been no consensus as to whether the information gained has been useful in predicting treatment failures or for adjusting therapy to target hypoxic radioresistance (Rasey et al. 1996; Rajendran et al. 2006). Until such time as tumor oxygenation can be routinely measured in cancer patients, it will be difficult to incorporate this information reliably into TCP modeling.

The difference in tumor cell intrinsic radiosensitivity between proliferating and quiescent clonogens, while real, is a factor of 2, at most (Chapter 3). Again, since the growth fraction of individual tumors is rarely measured and since its variation in groups of patients with similar cancers is not known, this tumor biology parameter is not currently available for incorporation into TCP models. However, it is likely that its effect will be much less than the large population variation in intrinsic radiosensitivities that has been reported for tumor cell lines and even less than the effect of tumor hypoxia. So while it would be welcomed if tumor prescriptions could be made for individual tumors whose biology parameters are known, radiotherapy research in the immediate future will continue to rely on information gained from *populations* of like tumors. And it should be again noted that since precise, individual values of the essential parameters for input into TCP models are not currently obtainable, best estimates or "guesstimates" will have to be made.

Radiation can also produce damages in the vasculature and stroma of solid tumors (Park et al. 2012) but it is not clear how these effects will factor into the LQ model of cell killing. These and immunological factors must also be taken into account if the clinical response of human cancer to radiotherapy is to be understood fully.

10.5 Future Perspectives

With regard to NTCP modeling (see Chapter 9; to be covered in detail in Volume II), much less is known of several of the limiting biological effects of interest. Where stem cell killing is responsible for the normal-tissue complication in question, the LQ model could play a useful role (Rutkowska, Baker and Nahum 2010; Rutkowska et al. 2012). But when vascular and stromal factors play a dominant causative role in the normal-tissue complication, it is difficult to see how the application of the LQ model can be sufficient. Consequently, the best approach at the moment is to continue to refine our data on normal-tissue complications from animal studies and from clinical data sets. While different mathematics for encoding the effects of tumor response and normal-tissue complications have been established by various groups (e.g., Nahum and Kutcher 2007; AAPM Task Group Report 2012), these models will only be as good as the basic radiobiology parameters that are input into them. And it is unfortunate that this type of research is receiving less funding under our present rules and priorities.

The Epilogue of this book identifies several research endeavors that the authors deem to be important for carrying TCP and NTCP modeling forward (Ruggieri and Nahum 2006; Nahum and Uzan 2012). Since many of the studies can be performed with classical radiobiology procedures, they might have to be integrated with molecular procedures that currently receive generous funding. Information gained from such basic understanding of fundamental radiobiology should only improve the usefulness of the proposed studies. As cancer treatment planning moves beyond the practice of fixed, uniform physical doses to PTVs towards "radiobiology-based planning," the whole research field should come together to assist with obtaining accurate values of those clonogen parameters that are most important for this objective. This textbook has attempted to collate the relevant quantitative information on these parameters currently available in the literature. Much more needs to be accomplished in order to improve the prescription of radiotherapy dose to cancer patients in general and to individual patients in particular.

Epilogue

In limiting our discussion of radiation mechanism mainly to research that utilized the LQ model, we may have disappointed some of our colleagues. We sincerely apologize but find that we do not have the time or skills to attempt to rationalize all of radiobiology. The advanced target theory that is presented in this textbook has not been proven by the experiments that are described. Nevertheless, there are several avenues of novel research that are suggested that could prove useful for extending quantitative radiobiology to improved treatment planning. Several of these directions for new laboratory research are now suggested.

For those that long for a different formulation to describe the radiation killing of tumor cells, the onus is on you to provide such a model and perform the laboratory research necessary to convince others of its validity. The LQ model has satisfied our needs to quantify radiation inactivation of both normal and tumor cells irradiated under most conditions of chemical and physical modulation. The most dominant chemical factor in dictating intrinsic radiosensitivity is the concentration of oxygen near the molecular target(s) at the time of irradiation. The most dominant physical factors are the type (quality) of radiation, its dose-rate and the temperature of cells during the radiation exposures. When all these factors are rigorously controlled, the vast majority of the over 3,000 survival curves that our research has generated can be described by unique values of α and β_o. When cells are irradiated at 37°C, repair of sublesions during the exposure must be taken into account.

Consequently, the rate of repair of sublethal damages generated by radiation in human tumor cells should be precisely measured to determine if it is similar to that reported for Chinese hamster lung fibroblasts. We suspect that they will be similar since the process of repair of "simple" lesions in DNA is an essential process for cell maintenance and survival (Hoeijmakers 2001; Clancy 2008). Any difference that exists between the rates of DNA sublethal damage repair between normal versus tumor cells would impact the effectiveness of optimal fractionated radiotherapy. These rates should be directly measured and reported.

Since the molecular lesions that produce radiation-induced α-inactivation are not repairable, it is likely that the current molecular biology investigations of DNA repair mechanisms will relate to β_o-inactivation and significantly impact the repair of normal tissues during fractionated protocols. How these repair pathways might be exploited for increasing the therapeutic ratio of current cancer treatments requires focused attention. But since the majority of cell killing by our current fractionated treatments comes from α-inactivation, new research should be directed to defining the chromatin target(s) involved, if these targets can be modulated for therapeutic gain and if they can be quantified in biopsy specimens to predict for the intrinsic radiosensitivity of individual tumors. This will require more studies performed with the dose fraction sizes utilized in clinical treatments.

The role of oxygen in the radiation killing of mammalian cells has been investigated for many years. As Chapter 6 describes, the different techniques devised for laboratory studies of tumor cell oxygenation might not produce equivalent concentrations of oxygen within different intracellular compartments. The suggestion that oxygen diffusion from the DNA targets of G_1-phase and mitotic cells follows kinetics characteristic of aqueous and lipid environments, respectively, requires follow-up research. If the DNA of chromosomes at mitosis has a lipid environment, radiation research could provide an answer for where the nuclear membrane lipid goes at this critical time at cell division. That it might be conserved during mitosis in association with chromosomal DNA suggests a more active role during the creation of the two daughter cells at cytokinesis than currently taught. It was radiobiology in the 1950s that provided the first estimates of viral DNA and RNA molecular weights through target theory calculations (Setlow and Pollard 1962). Radiation could still have an important role in researching the biophysics of structural elements in tumor and normal tissue cells.

It is our opinion that the conformation of DNA in cells at the various stages of the cell cycle will be important for determining their intrinsic radiosensitivity. As well, our research suggests that DNA conformation during the interphase of different tumor cells is not the same. We suggest that a much greater effort be placed on the molecular biology and biophysical research of chromatin in tumor cells with the intention of predicting their intrinsic radiosensitivity. The quantitative electron microscopy methods described in Chapter 6 are laborious but could be extended to a larger panel of human tumor cell lines and to additional molecular factors of importance. The tools of fluorescence microscopy available today are more than adequate for such studies. DNA conformation—particularly the folding and embedding of the 30-nm fibers in membrane—is likely to inform significantly about radiation sensitivity. And since histones appear to be important cell molecules that are chemically modified to bring about important changes in DNA structure in proliferating cells, these cellular proteins should take on a greater importance in understanding intrinsic cell radiosensitivity.

TCP and NTCP modeling requires improved input of significant tumor biology parameters. The $\bar{\alpha}$- and $\bar{\beta}_o$-inactivation parameters suggested for various human tumor pathologies (Table 7.2, Chapter 7) are reasonable starting values. These should be refined as more data are generated with cells irradiated under appropriate conditions. The number of clonogens in different human tumors requires appropriate estimates based upon knowledge of tumor volume and growth rate. And whether a portion of these clonogens are hypoxic at the time of radiotherapy will significantly impact the prediction of response. All these factors formed a field of research called "predictive assays" and are as important today as they were 25 years ago (Chapman, Peters and Withers 1989). And since our research field now enjoys many more technologies for investigations, new efforts to identify methods for obtaining these tumor biology parameters from individual human tumors should be intensified. It could be that, in the future, patient prescriptions will be defined by the $\bar{\alpha}$- and $\bar{\beta}_o$-parameters, the clonogen number, the growth fraction and the oxygenation status of each tumor. Until such time, the testing of novel fractionation schemes in groups of patients with similar cancer pathologies will indicate how much more can be gained from exploiting the improved imaging and dose delivery systems that are currently available.

If electron track-end lesions are such an important component of radiation dose that leads to cell death and tumor cures, novel physics research should attempt to increase their proportion in the dose delivered to planned treatment volumes (PTVs). It is instructive to consider from the data presented in this book that the spur events that lead to β-inactivation are persistent in all the various radiations that were investigated (except for the Bragg peaks of low-energy charged particles). How to deliver a larger proportion of electron track-end events to PTVs should command the attention of inventive radiation physicists for several years.

References

AAPM Task Group Report: The use of QA of biologically related models for treatment planning: Short report of the TG-166 of the therapy physics community of the AAPM. Li, X. A., Alber, M., Deasy, J. O., Jackson, A., Jee, K-W. K., Marks, L. B., Martel, M. K., Mayo, C., Moiseenko, V., Nahum, A., Niemierko, A., Semenenko, V. A. and Yorke, E. D. *Medical Physics* 39:1386–1409, 2012.

Adams, G. E. Radiation chemical mechanisms in radiation biology. *Advances in Radiation Chemistry* 3:125–208, 1972.

Alberts, B., Johnson, A., Lewis, J., Raff, M., Roberts, K. and Walter, P. *Molecular biology of the cell*, 4th Edition. New York: Garland Science, 2002.

Algan, O., Stobbe, C. C., Helt, A. M., Hanks, G. E. and Chapman, J. D. Radiation inactivation of human prostate cancer cells: The role of apoptosis. *Radiation Research* 146:267–275, 1996.

Alper, T. *Cellular radiobiology.* Cambridge, UK: Cambridge University Press, 1979.

Alper, T. and Howard-Flanders, P. The role of oxygen in modifying the radiosensitivity of *E. coli* B. *Nature, London* 178:978–979, 1956.

Baker F. and Sanger, I. The density of clonogenic cells in human tumors. *International Journal of Cell Cloning* 9:155–165, 1991.

Barendsen, G. W. Responses of cultured cells, tumors and normal tissues to radiations of different linear energy transfer. *Current Topics in Radiation Research* 4:293–356, 1968.

———. LET dependence of linear and quadratic terms in dose-response relationships for cellular damage: Correlations with the dimensions and structures of biological targets. *Radiation Protection Dosimetry* 31:384–387, 1990.

Barendsen, G. W. and Broerse, J. J. Experimental radiotherapy of a rat rhabdomyosarcoma with 16 MeV neutrons and 300 kVp x-rays: Effects of single exposures. *European Journal of Cancer* 5:373–391, 1969.

Barendsen, G. W. and Walter, H. M. D. Effects of different ionizing radiations on human cells in tissue culture: IV. Modification of radiation damage. *Radiation Research* 21:314–329, 1964.

Barendsen, G. W., Walter, H. M. D., Fowler, J. F. and Bewley, D. K., Effects of different ionizing radiations on human cells in tissue culture III. Experiments with cyclotron-accelerated alpha-particles and deuterons. *Radiation Research* 18:106–119, 1963.

Baumann, P., Nyman, J., Hoyer, M., Wennberg, B., Gagliardi, G., Lax, I., Drugge. N., Ekberg, L., Friesland S, Johansson, K-A., Lund, J. A., Morhed, E., Nilsson, K., Levin, N., Paludan, M., Sederholm, C., Traberg, A., Wittgren, L., Lewensohn, R. Outcome in a prospective phase II trial of medically inoperable stage I non-small-cell lung cancer patients treated with stereotactic body radiotherapy, *Journal of Clinical Oncology.*, 27:3290–3296, 2009.

Bedford, J. S. and Mitchell, J. B. Dose rate effects in synchronous mammalian cells in culture. *Radiation Research* 54:316–327, 1973.

Belli, M., Campa, A. and Ermolli, J. A semi-empirical approach to the evaluation of relative biological effectiveness of therapeutic proton beams: The methodological framework. *Radiation Research* 148:592–598, 1997.

Belli, M., Sapora, O. and Taboccini, M. A. Molecular targets in cellular response to ionizing radiation and implications for space radiation protection. *Journal of Radiation Research* 43 (Suppl) S13–S19, 2002.

Bentzen, S. M. Steepness of the clinical dose-control curve and variation in the *in vivo* radiosensitivity of head and neck squamous cell carcinoma. *International Journal of Radiation Biology* 61:417–423, 1992.

Bentzen, S. M., Thames, H. D. and Overgaard, J. Does variation in the *in vivo* cellular radiosensitivity explain the shallow clinical dose-control curve for malignant melanoma? *International Journal of Radiation Biology* 57:117–126, 1990.

Biade, S., Stobbe, C. C., Boyd, J. T. and Chapman, J. D. Chemical agents that promote chromatin compaction radiosensitize tumor cells. *International Journal of Radiation Biology* 77:1033–1042, 2001.

Biade, S., Stobbe, C. C. and Chapman, J. D. The intrinsic radiosensitivity of some human tumor cells throughout their cell cycles. *Radiation Research* 147:416–421, 1997.

Bidmead, M. and Rosenwald, J-C. Dose evaluation of treatment plans, in *Handbook of Radiotherapy Physics—Theory and practice* ed. P. Mayles, A. Nahum and J-C. Rosenwald, pp. 719–729. New York, London: Taylor & Francis, 2007.

Björk-Eriksson, T., West, C. M. L., Karlsson, E. and Mercke, C. Tumor radiosensitivity (SF_{2Gy}) is a prognostic factor for local control in head and neck cancers. *International Journal of Radiation Oncology Biology Physics* 46:13–19, 2000.

Blakely, E. A., Ngo, F. Q. H., Curtis, S. B. and Tobias, C. A. Heavy-ion radiobiology: Cellular studies. *Advances in Radiation Biology* 11:295–389, 1984.

Blakely, E. A., Tobias, C. A., Yang, C. H., Smith, K. C. and Lyman, J. T. Inactivation of human kidney cells by high-energy monoenergetic heavy-ion beams. *Radiation Research* 80:122–160, 1979.

Boag, J. W. In *Radiation research: Biomedical, chemical and physical perspectives,* ed. O. F. Nygaard, H. I. Adler and W. K. Sinclair, pp. 9–29. New York: Academic Press, 1975.

Branze, D. and Folani, M. Regulation of DNA repair throughout the cell cycle. *Nature Reviews Molecular Cell Biology* 9:277–308, 2008.

Brenner, D. J. and Hall, E. J. Fractionation and protraction for radiotherapy of prostate carcinoma. *International Journal of Radiation Oncology Biology Physics* 43:1095–1101, 1999.

Brock, W. A., Maor, M. H. and Peters, L. J. Cellular radiosensitivity as a predictor of tumor radiocurability. *Radiation Research* 104:290–296, 1985.

Brown, J. M. Evidence for acutely hypoxic cells in mouse tumors and a possible mechanism of reoxygenation. *British Journal of Radiology* 52:620–650, 1979.

Buffa, F. M., Davidson, S. E., Hunter, R. D., Nahum, A. E. and West, C. M. L. Incorporating biological measurements (SF_2, CFE) into a tumor control probability model increases their prognostic significance: A study in cervical carcinoma treated with radiation therapy. *International Journal of Radiation Oncology Biology Physics* 50:1113–1122, 2001.

Buffa, F. M., Fenwick, J. D. and Nahum, A. E. An analysis of the relationship between radiosensitivity and volume effects in tumor control probability modeling. *Medical Physics* 27:1258–1265, 2000.

Burki, H. J., Roots, R., Feinendegen, L. E. and Bond, V. P. Inactivation of mammalian cells after disintegration of ^3H and ^{125}I in cell DNA at −196°C. *International Journal of Radiation Biology* 24:363–375, 1973.

Buxton, G. V., Greenstock, C. L., Helman, W. B. and Ross, A. B. Critical review of rate constants for reactions of hydrated electrons, hydrogen atoms and hydroxyl radicals (.OH/.O⁻) in aqueous solution. *Journal of Physical and Chemical Reference Data* 17:513–886, 1988.

Chadwick, K. H. and Leenhouts, H. P. A molecular theory of cell survival. *Physics in Medicine and Biology* 18:78–87, 1973.

———. *The molecular theory of radiation biology.* Berlin: Springer-Verlag, 1981.

Chaplin, D. J., Durand, R. E. and Olive, P. L. Acute hypoxia in tumors: Implications for modifiers of radiation effects. *International Journal of Radiation Oncology Biology Physics* 12:1279–1282, 1986.

Chaplin, D. J., Olive, P. L. and Durand, R. E. Intermittent blood flow in a murine tumor: Radiobiological effects. *Cancer Research* 47:597–601, 1987.

Chapman, J. D. Biophysical models of mammalian cell inactivation by radiation. In *Radiation biology in cancer research,* ed. R. E. Meyn and H. R.Withers, pp. 21–32. New York: Raven Press, 1980.

———. Medical accelerator research institute in Alberta (MARIA). In *Pions and heavy ion radiotherapy: Preclinical and clinical studies,* ed. L. D. Skarsgard, pp. 37–46. New York: Elsevier Biomedical, 1983.

———. Tumor oxygenation. In *Encyclopedia of cancer,* vol. III, ed. J. Bertino, pp. 1914–1925. New York: Academic Press, 1997.

———. Single-hit mechanism of tumor cell killing by radiation. *International Journal of Radiation Biology* 79:71–81, 2003.

————. Can the two mechanisms of tumor cell killing by radiation be exploited for therapeutic gain? *Journal of Radiation Research* 55:2–9, 2014.

Chapman, J. D., Blakely, E. A., Smith, K. C. and Urtasun, R. C. Radiobiological characterization of the inactivating events produced in mammalian cells by helium and heavy ion. *International Journal of Radiation Oncology Biology Physics* 3:97–102, 1977.

Chapman, J. D., Blakely, E. A., Smith, K. C., Urtasun, R. C., Lyman, J. T. and Tobias, C. A. Radiation biophysical studies with mammalian cells and a modulated carbon ion beam. *Radiation Research* 101–111, 1978.

Chapman, J. D., Doern, S. D., Reuvers, A. P., Gillespie, C. J., Chatterjee, A., Blakely, E. A., Smith, K. C. and Tobias, C. A. Radioprotection by DMSO of mammalian cells exposed to x-rays and heavy charged-particle beams. *Radiation and Environmental Biophysics* 16:29–41, 1979.

Chapman, J. D., Dugle, D. L., Reuvers, A. P., Meeker, B. E. and Borsa, J. Studies on the radiosensitizing effect of oxygen in Chinese hamster cells. *International Journal of Radiation Biology* 26:383–389, 1974.

Chapman, J. D., Franko, A. and Sharplin, J. A marker for hypoxic cells in tumors with potential clinical applicability. *British Journal of Cancer* 43:546–550, 1981.

Chapman, J. D. and Gillespie, C. J. Radiation-induced events and their time-scale in mammalian cells. *Advances in Radiation Biology* 9:143–198, 1981.

————. The power of radiation biophysics—Let's use it. *International Journal of Radiation Oncology Biology Physics* 84:309–311, 2012.

Chapman, J. D., Gillespie, C. J., Reuvers, A. P. and Dugle, D. L., The inactivation of Chinese hamster cells by x-rays: The effects of chemical modifiers on single- and double-events. *Radiation Research* 64:365–375, 1975.

Chapman, J. D., Peters, L. J. and Withers, H. R. *Prediction of tumor treatment response.* Oxford: Pergamon Press, 1989.

Chapman, J. D. and Reuvers, A. P. The time-scale of radioprotection in mammalian cells. In *Radioprotection, Chemical Compounds, Biological Means,* ed. A. Locker and K. Flemming, pp. 9–18. Basel: Birkhaeuser, 1977.

Chapman, J. D., Reuvers, A. P., Borsa, J. and Greenstock, C. L. Chemical radioprotection and radiosensitization of mammalian cells growing *in vitro. Radiation Research* 56:291–306, 1973.

Chapman, J. D., Reuvers, A. P., Doern, S. D., Gillespie, C. J. and Dugle, D. L. Radiation chemical probes in the study of mammalian cell inactivation and their influence on radiobiological effectiveness. In *Proceedings of Fifth Symposium on Microdosimetry,* ed. J. Booz, H. G. Ebert and B. G. R. Smith, pp. 775–798. Luxembourg: Commission of the European Communities, 1976.

Chapman, J. D., Schneider, R. F., Urbain, J-L. and Hanks, G. E. Single-photon emission computed tomography and positron-emission tomography assays for tissue oxygenation. *Seminars in Radiation Oncology* 11:47–57, 2001.

Chapman, J. D., Stobbe, C. C., Arnfield, M. R., Santus, R., Lee, J. and McPhee, M. S. Oxygen dependency of tumor cell killing *in vitro* by light-activated Photofrin II. *Radiation Research* 126:73–79, 1991.

Chapman, J. D., Stobbe, C. C., Gales, T., Das, I. J., Zellmer, D. L., Biade, S. and Matsumoto, Y. Condensed chromatin and cell inactivation by single-hit kinetics. *Radiation Research* 151:433–441, 1999.

Chapman, J. D., Stobbe, C. C. and Matsumoto, Y. Chromatin compaction and tumor cell radiosensitivity at 2 Gy. *American Journal of Clinical Oncology* 24:509–515, 2001.

Chapman, J. D., Sturrock, J., Boag, J. W. and Crookall, J. O. Factors affecting the oxygen tension around cells growing in plastic petri dishes. *International Journal of Radiation Biology* 17:305–328, 1970.

Chapman, J. D., Todd, P. and Sturrock, J. X-ray survival of cultured Chinese hamster cells resuming growth after plateau phase. *Radiation Research* 42:590–600, 1970.

Chapman, J. D., Urtasun, R. C., Blakely, E. E., Smith, K. C. and Tobias, C. A. Hypoxic cell sensitizers and heavy charged-particle radiations. *British Journal of Cancer* 37 (Suppl. III):184–188, 1978.

Chatterjee, A. and Holley, W. Computer simulations of initial events in the bio-chemical mechanisms of DNA damage. *Advances in Radiation Biology* 17:181–226, 1993.

Clancy, S. DNA damages and repair: Mechanisms for maintaining DNA integrity. *Nature Education* 1:103, 2008.

Cole, A., Cooper, W. G., Shonka, F., Corry, P. M., Humphrey, R. M. and Ansevin, A. T. DNA scission in hamster cells and isolated nuclei studied by low-voltage electron beam radiation. *Radiation Research* 60:1–33, 1974.

Cole, A., Shonka, F., Corry, P. and Cooper, W. G. CHO cell repair and double-strand DNA breaks induced by gamma and alpha radiations, in *Molecular mechanisms for repair of DNA*, ed. R. B. Setlow and P. C. Hanawalt, pp. 665–676. New York: Plenum, 1975.

Costello, J. F., Berger, M. S., Huang, H. S. and Cavenee, W. K. Silencing of p16/CDKN2 expression in human gliomas by methylation and chromatin con-densation. *Cancer Research* 56:2405–2410, 1996.

Courtney, V. D. and Mills, J. An *in vitro* assay for human tumors grown in immune-suppressed mice and treated *in vivo* with cytotoxic agents. *British Journal of Cancer* 37:261–268, 1978.

Coutard, H. Die Röntgenbehandlung der Epithelialen Krebse der Tonsillengegend. *Strahlentherapie* 33:249–252, 1929.

Culo, F., Yuhas, J. M. and Ladman, A. J. Multicellular tumor spheroids: A model for combined *in vivo/in vitro* assay of tumor immunity. *British Journal of Cancer* 41:100–112, 1980.

Curtis, S. B. Lethal and potentially lethal lesions induced by radiation: A unified repair model. *Radiation Research* 106:252–270, 1986.

Dale, R. G. Radiobiology of brachytherapy, in *Handbook of radiotherapy physics—Theory and practice,* ed. P. Mayles, A. Nahum and J-C. Rosenwald, pp. 1181–1199, New York, London: Taylor & Francis, 2007.

Das, I. J. and Chopra, K. L. Backscatter dose perturbation in kilo-voltage pho-ton beams at high atomic number interfaces. *Medical Physics* 22:767–773, 1995.

Dasika, G. K., Lin, S-C., J., Zhao, S., Sung, P., Tomkinson A. and Lee, E. Y-H. DNA damage-induced cell cycle checkpoints and DNA strand break repair in development and tumorigenesis. *Oncogene* 55:7883–7899, 1999.

Daşu, A., Toma-Daşu, I. and Fowler, J. F. Should single or distributed parameters be used to explain the steepness of tumor control probability curves? *Physics in Medicine and Biology*, 48:387–397, 2003.

Deacon, J., Peckham, M. J. and Steel, G. G. The radioresponsiveness of human tumors and the initial slope of the cell survival curve. *Radiotherapy and Oncology* 2:317–323, 1984.

DeNardo, G. L., Schlom, J., Bushsbaum, D. J., Meredith, R. F., O'Donoghue, J. A., Humm, J. and DeNardo, S. J. Rationales, evidence and design considerations for fractionated radioimmunotherapy. *Cancer* 94:1332–1348, 2002.

Dingerfelder, M. Track-structure simulations for charged particles. *Health Physics* 103:590–595, 2012.

Dugle, D. L. and Gillespie, C. J. In *Molecular mechanisms for repair of DNA,* ed. P. C. Hanawalt and R. B. Setlow, pp. 685–687. New York: Plenum Press, 1975.

Dugle, D. L., Gillespie, C. J. and Chapman, J. D. DNA strand breaks, repair and survival in x-irradiated mammalian cells. *Proceedings of National Academy of Sciences* 73:809–812, 1976.

Durand, R. E. and Sutherland, R. M. Effects of intercellular contact on repair of radiation damage. *Experimental Cell Research* 71:75–80, 1972.

———. Growth and radiation survival characteristics of V79-171b Chinese hamster cells: A possible influence of intercellular contact. *Radiation Research* 56:513–527, 1973a.

———. Dependence of the radiation response of an *in vitro* tumor model on cell cycle effects. *Cancer Research* 33:213–219, 1973b.

Elkind, M. M. Sedimentation of DNA released from Chinese hamster cells. *Biophysics Journal* 11:502–520, 1971.

Elkind, M. M., Han, A. and Volz, K. W. Radiation response of mammalian cells grown in culture. IV. Dose-dependence of division delay and post-irradiation growth of surviving and non-surviving Chinese hamster cells. *Journal of National Cancer Institute* 30:705–721, 1963.

Elkind, M. M., Sutton-Gilbert, H., Moses, W. B., Alescio, T. and Swain, R. B. Radiation response of mammalian cells in culture. V. Temperature dependence of the repair of x-ray damage in surviving cells (aerobic and hypoxic). *Radiation Research* 25:359–376, 1965.

Elkind, M. M., Swain, R. W., Alescio, T., Sutton, H. and Moses, W. B. Oxygen, nitrogen, recovery and radiation therapy. In *Cellular radiation biology,* pp. 442–461. Baltimore: Williams and Wilkins, 1965.

Elkind, M. M. and Whitmore, G. F. *The radiobiology of cultured cells.* New York: Gordon and Breach, 1967.

Elsässer, T., Weyrather, W. K., Friedrich, T., Durante, M., Iancu, G., Krämer, M., Kragl, G., Brons, S., Winter, M., Weber, K-J. and Scholtz, M. Quantification of the relative biological effectiveness for ion beam radiotherapy.

Direct experimental comparison of proton and carbon ion beams and a novel approach for treatment planning. *International Journal of Radiation Oncology Biology Physics* 78:1177–1183, 2010.

Emami, B., Lyman, J., Brown, A., Coia, L., Goitein, M., Munzenrider, J. E., Shank, B., Solin, L. J. and Wesson, M. Tolerance of normal tissue to therapeutic radiation. *International Journal of Radiation Oncology Biology Physics* 21:109–122, 1991.

Emfietzoglou, D., Papamichael, G., Kostarelos, K. and Moscovitch, M. A Monte Carlo track structure code for electrons (approximately 10 eV–10 keV) and protons (approximately 0.3–10 MeV) in water: Partitioning of energy and collision events. *Physics in Medicine and Biology* 45:3171–3194, 2000.

Fertil, B. and Malaise, E. P. Inherent radiosensitivity as a basic concept for human tumor radiotherapy. *International Journal of Radiation Oncology Biology Physics* 7:621–629, 1981.

Fowler, J. F. Half-times of irradiation recovery in accelerated partial breast irradiation: Incomplete recovery as a potentially dangerous enhancer of radiation damage. *Journal of Cancer Research & Therapy* 1:230–234, 2013.

Fowler, J. F. The linear-quadratic formula and progress in fractionated radiotherapy. *British Journal of Radiology* 62:679–694, 1989.

Frese, M. C., Yu, V. K., Stewart, R. D. and Carlson, D. J. A mechanism-based approach to predict the relative biological effectiveness of protons and carbon ions in radiation therapy. *International Journal of Radiation Oncology Biology Physics* 83:442–450, 2012.

Gillespie, C. J., Chapman, J. D., Reuvers, A. P. and Dugle, D. L. The inactivation of Chinese hamster cells by x-rays: Synchronized and exponential cell populations. *Radiation Research* 64:353–364, 1975.

Gillespie, C. J., Dugle, D. L., Chapman, J. D., Reuvers, A. P. and Doern, S. D. DNA damage and repair in relation to mammalian cell survival: Implications for microdosimetry. In *Proceedings of Fifth Symposium on Microdosimetry,* ed. J. Booz, H. G. Ebert and B. G. R. Smith, pp. 799–813. Luxembourg: Commission of the European Communities, 1976.

Goitein, M. The comparison of treatment plans. *Seminars in Radiation Oncology* 2:246–256, 1992.

Gonzalez, S., Arnfield, M. R., Meeker, B. E., Tulip, J., Lakey, W. H., Chapman, J. D. and McPhee, M. S. Treatment of Dunning R3327-AT rat prostate tumors with photodynamic therapy in combination with misonidazole. *Cancer Research* 46:2858–2862, 1986.

Goodhead, D. T. Initial events in the cellular effects of ionizing radiations: Clustered damage in DNA. *International Journal of Radiation Biology* 65:7–17, 1994.

———. Saturable repair models of radiation action in mammalian cells. *Radiation Research* 104:558–567, 1985.

Gray, L. H., Conger, A. D., Ebert, M., Hornsey, S. and Scott, O. C. A. The concentration of oxygen dissolved in tissues at the time of irradiation as a factor in radiotherapy. *British Journal of Radiology* 26:638–648, 1953.

Greenstock, C. L., Chapman, J. D., Raleigh, J. A., Shierman, E. and Reuvers, A. P. Competitive radioprotection and radiosensitization in chemical systems. *Radiation Research* 59:556–571, 1974.

Groshar, D., McEwan, J. B., Parliament, M. B, Urtasun, R. C., Golberg, L. E., Hoskinson, M., Mercer, J. R., Mannan, R. H., Wiebe, L. I. and Chapman, J. D. Imaging tumor hypoxia and tumor perfusion. *Journal of Nuclear Medicine* 34:885–888, 1993.

Gurley, L. R., Tobey, R. A., Walters, R. A., Hildebrand, C. E., Hohmann, P. G., D'Anna, J. A., Barham, S. S. and Deavan, L. L. Histone phosphorylation and chromatin structure in synchronized mammalian cells. In *Cell cycle regulation*, ed. J. Jeter, I. L. Cameron, G. M. Padilla and A. M. Zimmerman, pp. 37–60. New York: Academic Press, 1978.

Hahn, G. M and Little, B. Plateau-phase cultures of mammalian cells: An *in vitro* model for human cancer. *Current Topics in Radiation Research* 8:39–83, 1972.

Hahn, G. M., Stewart, H. R., Yang, S.-J. and Parker, V. Chinese hamster cell monolayer cultures. I. Changes in cell dynamics and modifications of the cell growth cycle with the period of growth. *Experimental Cell Research* 49:285–292, 1968.

Hall, E. J. *Radiobiology for the radiologist.* Philadelphia: Lippincott Williams & Wilkins, 2000.

Hall, E. J. and Giaccia, A. J. *Radiobiology for the radiologist.* Philadelphia: Lippincott Williams & Wilkins, 2006.

Hanks, G. E., Schultheiss, T. E., Hanlon, A. L., Hunt, M., Lee, W. R. and Epstein, B. E. Optimization of conformal radiation treatment of prostate cancer: Report of a dose escalation study. *International Journal of Radiation Oncology Biology Physics* 37:543–550, 1997.

Herold, D. M., Das, I. J., Stobbe, C. C., Iyer, R. V. and Chapman, J. D. Gold microspheres: A selective technique for producing biologically effective dose enhancement. *International Journal of Radiation Biology* 76:1357–1364, 2000.

Hewitt, H. B. and Wilson, C. W. Survival curves for tumor cells irradiated *in vivo*. *Annals of New York Academy of Sciences* 95:818–827, 1961.

Hill, M. A. The variation in biological effectiveness of x-rays and gamma rays with energy. *Radiation Protection Dosimetry* 53:233–244, 2008.

Hill, R. P. and Bush. R. S. A lung colony assay to determine the radiosensitivity of cells of a solid tumor. *International Journal of Radiation Biology* 15:435–444, 1969.

Hoeijmakers, J. H. J. Genome maintenance mechanisms for preventing cancer. *Nature* 411:366–374, 2001.

Hoffmann, A. L. and Nahum, A. E. Fractionation in normal tissues: The $(\alpha/\beta)_{eff}$ concept can account for dose heterogeneity and volume effects. *Physics in Medicine and Biology* 58:6897–6914, 2013.

Howard, A. and Pelc, S. Synthesis of desoxyribonucleic acid in normal and irradiated cells and its relation to chromosome breakage. *Heredity Suppl.* 6:261–273, 1953.

ICRU (International Commission on Radiation Units and Measurements), report 62. Prescribing, recording and reporting photon beam therapy. Supplement to ICRU report 50, ICRU, Bethesda, MD, 1999.

Iliakis, G., Wang, Y., Guan, J. and Wang, H. DNA damage check point control in cells exposed to ionizing radiation. *Oncogene* 22:5834–5847, 2003.

Jaggi, B. and Palcic, B. The design and development of an optical scanner for cell biology. *IEEE 7th Engineering in Medicine and Biology* 2:890–985, 1985.

Jensen, E. V. and DeSombre, E. R. Estrogen-receptor interactions. *Science* 182:126–134, 1973.

Johns, H. E. and Cunningham, J. R. *The physics of radiology,* 4th ed. Springfield, IL: Charles C Thomas, 1983.

Joiner, M. C., Marples, B., Lambin, P., Short, S. C. and Turesson, I. Low-dose hypersensitivity: Current status and possible mechanisms. *International Journal of Radiation Oncology Biology Physics* 49:379–389, 2001.

Kagawa, K., Murakami, M., Hishikawa, Y., Abe, M., Akagi, T., Yonou, T., Kagiya, G., Furusawa, Y., Ando, K., Nojima, K., Aoki, M. and Kanai, T. Preclinical biological assessment of proton and carbon ion beams at Hyogo beam medical center. *International Journal of Radiation Oncology Biology Physics* 54:928–938, 2002.

Källman, P., Ågren, A. and Brahme, A. Tumour and normal tissue responses to fractionated non-uniform dose delivery. *International Journal of Radiation Biology* 62:249–262, 1992.

Kanai, T., Endo, M., Minohara, S., Miyahara, N. and Kryama-Ito, H. Biophysical characterization of HIMAC clinical irradiation system for heavy-ion radiation therapy. *International Journal of Radiation Oncology Biology Physics* 44:201–210, 1999.

Kassis, A. I. The amazing world of Auger electrons. *International Journal of Radiation Biology* 80:789–803, 2004.

———. Therapeutic radionuclides: Biophysical and radiobiologic principles. *Seminars in Nuclear Medicine* 38:358–366, 2008.

Kauffman, M. G., Noga, S. J., Kelly, T. J. and Donnenberg, A. D. Isolation of cell cycle fraction by counter flow centrifugal elutriation. *Analytical Biochemistry* 191:41–46, 1990.

Kaufmann, W. K., Cell cycle checkpoints and DNA repair preserve the stability of the human genome. *Cancer Metastasis Reviews* 1:31–41, 1995.

Kellerer, A. M. and Rossi, H. H. The theory of dual radiation action. *Current Topics in Radiation Research* 8:85–158, 1972.

Kerr, J. F. R. and Searle, J. Apoptosis: Its nature and kinetic role. In *Radiation biology and cancer research,* ed. R. E. Meyn and H. R. Withers, pp. 367–384, New York: Raven Press, 1980.

Kirkpatrick, J. P., Meyer, J. J. and Marks, L. B. The linear-quadratic model is inappropriate to model high dose per fraction effects in radiotherapy. *Seminars in Radiation Oncology* 18:240–243, 2008.

Kirkpatrick, J. P., van der Kogel, A. J. and Schultheiss, T. E. Radiation dose-volume effects in the spinal cord. *International Journal of Radiation Oncology Biology Physics* 76 (no. 3, supplement): S42–S49, 2010.

Koch, C. J., Kruuv, J. and Frey, H. E. Variation in radiation response as a function of oxygen tension. *Radiation Research* 53:33–42, 1973.

Koch, C. J. and Painter, R. B. The effect of extreme hypoxia on the repair of DNA single-strand breaks in mammalian cells. *Radiation Research* 64:256–269, 1975.

Kooi, M. W., Stap, J. and Barendsen, G. W. Proliferation kinetics of cultured cells after irradiation with x-rays and 14 MeV neutrons studied by time-lapse cinematography. *International Journal of Radiation Biology and Related Studies in Physics Chemistry and Medicine* 45:583–592, 1984.

Kornberg, R. D. Chromatin structure: A repeating unit of histones and DNA. Chromatin structure based on a repeating unit of eight histone molecules and about 200 base pairs of DNA. *Science* 184:868–871, 1974.

Kornberg, R. D. and Klug, A. The nucleosome. *Scientific American* 244 (2): 52–64, 1981.

Kraft, G. Tumor therapy with heavy charged particles. *Progress in Particle and Nuclear Physics* 45 (Suppl. 2): S473–S544, 2000.

Krämer, M. and Kraft, G. Calculations of heavy-ion track structure. *Radiation and Environmental Biophysics* 33:91–109, 1994.

Krämer, M. and Scholz, M. Treatment planning for heavy ion radiotherapy: Calculation and optimization of biologically effective dose. *Physics in Medicine and Biology* 45:3319–3330, 2000.

Kutcher, G. J. and Burman, C. Calculation of complication probability factors for non-uniform normal tissue irradiation. *International Journal of Radiation Oncology Biology Physics* 16:1623–1630, 1989.

Kutcher, G. J., Burman, C., Brewster, L., Goitein, M. and Mohan, R. Histogram reduction method for calculating complication probabilities for three-dimensional treatment planning evaluations. *International Journal of Radiation Oncology Biology Physics* 21:137–146, 1991.

Lange, C. S., Cole, A. and Ostashevsky, J. Y. Radiation-induced damage to chromosomal DNA molecules: Deduction of chromosomal DNA organization from the hydrodynamic data used to measure DNA double-strand breaks and from stereo electron microscopic observations. *Advances in Radiation Biology* 17:261–421, 1993.

Lea, D. E. *Action of radiations on living cells.* Cambridge: University Press, 1962.

Lechtman, E., Chattopadhyay, N., Cal, Z., Mashouf, S., Reilly, R. and Pignot, J. P. Implications on clinical scenario of gold nanoparticle radiosensitization in regards to photon energy, nanoparticle size, concentration and location. *Physics in Medicine and Biology* 56:4631–4647, 2011.

Ling, C. C., Gerweck, L. E., Zaider, M. and Yorke, E. Dose-rate effects in external beam radiotherapy redux. *Radiotherapy and Oncology* 95:261–268, 2010.

Lyman, J. T. Complication probability as assessed from dose volume histograms. *Radiation Research* (Supplement) 8:S13–S19, 1985.

Mador, D., Ritchie, B., Meeker, B., Moore, R., Elliot, F.G., McPhee, M. S., Chapman, J. D. and Lakey, W. H. Response of the Dunning R3327-H prostate adenocarcinoma to radiation and various chemotherapeutic agents. *Cancer Treatment Reports* 66:1837–1843, 1982.

Marin, G. and Bender, M. A. Survival kinetics of HeLa S-3 cells after incorporation of ^{3}H-thymidine and ^{3}H-uridine. *International Journal of Radiation Biology* 7:221–234, 1963a.

———. A comparison of mammalian cell killing by incorporated ^{3}H-thymidine and ^{3}H-uridine. *International Journal of Radiation Biology* 7:235–244, 1963b.

Marks, L. B., Bentzen, S. M., Deasy, J. O., Kong, F-M. S., et al. Radiation dose-volume effects in the lung. *International Journal of Radiation Oncology Biology Physics* 76 (no. 3, supplement): S70–S76, 2010.

Marks, L. B., Ten Haken, R. T. and Martel, M. (eds.). Quantitative analyses of normal tissue effects in the clinic (QUANTEC), *International Journal of Radiation Oncology Biology Physics* 76 (no. 3, supplement): 2010.

Marks, L. B., Yorke, E. D., Jackson, A., Ten Haken, R. K., et al. Use of normal tissue complication probability models in the clinic. *International Journal of Radiation Oncology Biology Physics* 76 (no. 3, supplement): S10–S19, 2010.

Marples, B. and Collis, S. J. Low-dose hyper-radiosensitivity: Past, present and future. *International Journal of Radiation Oncology Biology Physics* 70:1310–1318, 2008.

Matheson, I. B. C. and Rodgers, M. A. J. Oxygen transfer rates across water–micelle interfaces derived from measurements of Ni^{2+} quenching of singlet molecular oxygen in aerosol-OT-heptane reverse micelles. *Journal of Physical Chemistry* 86:884–887, 1982.

Mayles, P., Nahum, A. and Rosenwald, J-C. *Handbook of radiotherapy physics—Theory and practice.* New York: Taylor & Francis, 2007.

Mayles, P. and Williams, P. Megavoltage photon beams, ch. 22, in *Handbook of radiotherapy physics—Theory and practice*, ed. P Mayles, A. Nahum and J-C. Rosenwald, pp. 451–481, New York, London: Taylor & Francis, 2007.

Mayo, C., Martel, M. K., Marks, L. B., Flickinger, J., Nam, J. and Kirkpatrick, J. Radiation dose-volume effects of optic nerves and chiasm. *International Journal of Radiation Oncology Biology Physics* 76 (no. 3, supplement): S28–S35, 2010.

McMahon, S. J., Mendenhall, M. H., Jain, S. and Currell, F. Radiotherapy in the presence of contrast agents: A general figure of merit and its application to gold nanoparticles. *Physics in Medicine and Biology* 53:5635–5651, 2008.

Michaud, W. A. and Sanche, L. Cross sections for low-energy (1–100 eV) electron elastic and inelastic scattering in amorphous ice. *Radiation Research* 159:3–22, 2003.

Milosevic, M., Warde P., Ménard C., Chung, P., Toi, A., Ishkanian, A., McLean, M., Pinille, M., Sykes, J., Gospoderowicz, M., Cattan, C., Hill, R. P. and Bristow, R. Tumor hypoxia predicts biochemical failure following radiotherapy for clinically localized prostate cancer. *Clinical Cancer Research* 18:2108–2114, 2012.

Moiseenko, V. V., Hamm, R. N., Waker, A. J. and Prestwich, W. V., Modeling DNA damage induced by different energy photons and tritium beta-particles. *International Journal of Radiation Biology* 74:533–550, 1998.

Moore, R. B., Chapman, J. D., Mercer, J. R., Mannan, R. H., Wiebe, L. I., McEwan, A. J. and McPhee, M. S. Measurement of PDT-induced hypoxia in Dunning prostate tumors by iodine-123-iodoazomycin arabinoside. *Journal of Nuclear Medicine* 34:405–413, 1993.

Moore R. B., Chapman, J. D., Mokrzanowski, A. D., Arnfield, M. R., McPhee, M. S. and McEwan, A. J. Non-invasive monitoring of photodynamic therapy with ^{99}Technetium HMPAO scintigraphy. *British Journal of Cancer* 65:491–497, 1992.

Movsas, B., Chapman, J. D., Greenberg, R. E., Hanlon, A. L., Horwitz, E. M., Pinover, W. H., Stobbe, C. and Hanks, G. E. Increasing levels of hypoxia in prostate carcinoma correlate significantly with increasing clinical stage and patient age. An Eppendorf pO_2 study. *Cancer* 89:2018–2024, 2000.

Movsas, B., Chapman, J. D., Hanlon, A. L., Horwitz, E. M., Greenberg, R. E., Stobbe, C., Hanks, G. E. and Pollack, A. Hypoxic prostate/muscle pO_2 ratio predicts for biochemical failure in patients with prostate cancer: Preliminary findings. *Urology* 60:634–639, 2002.

Mozumder, A. and Magee, J. L. Model of tracks of ionizing radiations for radical reaction mechanisms. *Radiation Research* 28:203–214, 1966a.

———. A simplified approach to diffusion-controlled radical reactions in the tracks of ionizing particles. *Radiation Research* 28:215–231, 1966b.

Muller, S. and Wienberg, J. Multicolor chromosome bar codes. *Cytogenetic Genome Research* 114:245–249, 2006.

Munroe, T. R. The relative radiosensitivity of the nucleus and cytoplasm in Chinese hamster fibroblasts. *Radiation Research* 42:451–470, 1970.

Nahum, A. and Kutcher, G. Biological evaluation of treatment plans, ch. 36, in *Handbook of radiotherapy physics—Theory and practice,* ed. P. Mayles, A. Nahum and J-C. Rosenwald, pp. 731–771. New York, London: Taylor & Francis, 2007.

Nahum, A. E., Movsas, B., Horwitz E. M., Stobbe, C. C. and Chapman J. D., Incorporating clinical measurements of hypoxia into tumor local control modeling of prostate cancer: Implications for the α/β ratio. *International Journal of Radiation Oncology Biology Physics* 57:391–401, 2003.

Nahum, A. E. and Sanchez-Nieto, B. Tumor control probability modeling: Basic principles and applications in treatment planning. *Physica Medica* 17:13–23, 2001.

Nahum, A. E. and Tait, D. M. Maximizing local control by customized dose prescription for pelvic tumors: Tumor response modeling and treatment planning. In *Advanced radiation therapy: Tumor response monitoring and treatment planning,* ed. A. Breit, pp. 425–431, Berlin: Springer-Verlag, 1992.

Nahum, A. E. and Uzan, J. (Radio)biological optimization of external-beam radiotherapy. *Computational and Mathematical Methods in Medicine,* vol. 2012, Article ID 329214, 13 pages, 2012. doi:10.1155/2012/329214.

Niemierko, A. A generalized concept of equivalent uniform dose (EUD). *Medical Physics* 26:1101, 1999.

Nikjoo, H. Radiation track and DNA damage. *Iran Journal of Radiation Biology* 1:3–16, 2003.

Nikjoo, H., Emfietzoglou, D. and Charlton, D. E. The auger effect in physical and biological research. *International Journal of Radiation Biology* 84:1011–1026, 2008.

Nurse, P. Universal control mechanism regulating onset of M-phase. *Nature* 344:503–508, 1990.

Okada, S. *Radiation biochemistry. Volume I: Cells.* New York: Academic Press, 1970.

Olive, P. L. Molecular approaches to detecting DNA damage. In *DNA damage and repair*, vol. 2, ed. J. A. Nickoloff and M. F. Hoekstra, pp. 539–557. Totowa, NJ: Humana Press, 2000.

Olive, P. L. and Banath, J. P. Detection of DNA double-strand breaks through the cell cycle after exposure to x-rays, bleomycin, etoposide and [125]IUrd. *International Journal of Radiation Biology* 64:349–358, 1993.

Olive, P. L., Banath, J. P. and Durand, R. E. Heterogeneity in radiation-induced DNA damage and repair in tumor and normal cells measured using the "comet" assay. *Radiation Research* 122:69–72, 1990.

Ostling, O. and Johanson, K. J. Microelectrophoretic study of radiation-induced DNA damages in individual mammalian cells. *Biochemical and Biophysical Research Communications* 123:291–298, 1984.

Overgaard J. and Horsman M. R. Modification of hypoxia-induced radioresistance in tumors by the use of oxygen and sensitizers. *Seminars in Radiation Oncology* 6:10–21, 1996.

Paganetti, H., Niemierko, A., Ancukiewicz, M., Gerweck, L. E., Goitein, M., Loeffler, J. S. and Suit, H. D. Relative biological effectiveness (RBE) values for proton beam therapy. *International Journal of Radiation Oncology Biology Physics* 53:407–421, 2002.

Palcic, B., Poon, S. S. S., Thurston, G. and Jaggi, B. Time-lapse records of cells *in vitro* using optical memory disk and cell analyzer. *Journal of Tissue Culture Methods* 11:19–22, 1988.

Palcic, B. and Skarsgard, L. D. Reduced oxygen enhancement ratio at low doses of ionizing radiation. *Radiation Research* 100:328–329, 1984.

Park, C., Papiez, L., Zhang, S., Story, M. and Timmerman, R. D. Universal survival curve and single fraction equivalent dose: Useful tools in understanding potency of ablative radiotherapy. *International Journal of Radiation Oncology Biology Physics* 70:847–852, 2008.

Park, H. J., Griffin, R. J., Hui, S., Levitt, S. H. and Song, C. W. Radiation-induced vascular damage in tumors: Implications of vascular damage in ablative hypofractionated radiotherapy (SBRT and SRS). *Radiation Research* 177:311–327, 2012.

Parliament, M. B., Chapman, J. D., Urtasun, R. C., McEwan, A. J., Goldberg, L., Mercer, J. R., Mannan, R. H. and Wiebe, L. I. Non-invasive assessment of human tumor hypoxia with ^{123}I-iodoazomycin arabinoside: Preliminary report of a clinical study. *British Journal of Cancer* 65:90–95, 1992.

Paull, T. T., Rogakou, E. P., Yamazaki, V., Kirchgessner, C. U., Gellert, M. and Bonner, W. M. A critical role for histone H2AX in recruitment of repair factors to nuclear foci after DNA damage. *Current Biology* 10–886–895, 2000.

Peacock, J. H., Eady, J. J., Edwards, S. M., McMillan, T. J. and Steel, G. G. The intrinsic α/β ratio for human tumor cells: Is it constant? *International Journal of Radiation Biology* 61:479–487, 1992.

Pearson P. The use of new staining techniques for human chromosome identification. *Journal of Medical Genetics* 9:264–275, 1972.

Plante, I., Ponomarev, A. and Cucinotta, F. A. 3D visualization of the stochastic patterns of the radial dose in nano-volumes by a Monte Carlo simulation of HZE ion track structure. *Radiation Protection Dosimetry* 143:156–161, 2011.

Pollard, J. M. and Gatti, R. A. Clinical radiation sensitivity with DNA repair disorders. *International Journal of Radiation Oncology Biology Physics* 74:1323–1331, 2009.

Pollard, K. J., Samuels, M. L., Crowley, K. A., Hansen, J. C. and Peterson, C. L. Functional interaction between GCN5 and polyamines: A new role for core histone acetylation. *EMBO Journal* 18:5622–5633, 1999.

Price, W. A., Stobbe, C. C., Park, S. J. and Chapman, J. D. Radiosensitization of tumor cells by cantharidin and some analogues. *International Journal of Radiation Biology* 80:269–279, 2004.

Prise, K. M. Studies of bystander effects in human fibroblasts using a charged particle microbeam. *International Journal of Radiation Biology* 74:793–798, 1998.

Radford, I. R. The level of induced DNA double-strand breakage correlates with cell killing after x-radiation. *International Journal of Radiation Biology* 48:45–54, 1985.

Rajendran, J. G., Schwartz, D. L., O'Sullivan, J., Peterson, L. M., Scharnhorst, J., Grierson, J. R. and Krohn, K. A. Tumor hypoxia imaging with [F-18] fluoromisonidazole positron emission tomography in head and neck cancer. *Clinical Cancer Research* 12:5435–5441, 2006.

Raleigh, J. A., Franko, A., Kelly, D., Trimble, L. and Allan, P. Development of an *in vivo* [19]FMR method for measuring oxygen deficiency in tumors. *Magnetic Resonance Medicine* 22:451–466, 1991.

Raleigh, J. A., Miller, G. G. and Franko, A. J. Fluorescence immunohistochemical detection of hypoxic cells in spheroids and tumors. *British Journal of Cancer* 56:395–400, 1987.

Rasey, J. S., Koh, W. J., Evans, M. L., Peterson, L. M., Lewellen, T. K., Graham, M. M. and Krohn, K. A. Quantifying regional hypoxia in human tumors with positron emission tomography of [^{18}F] fluoromisonidazole: A pretherapy study of 37 patients. *International Journal of Radiation Oncology Biology Physics* 36:417–428, 1996.

Rauth, A. M. and Simpson, J. A. The energy loss of electrons in solids. *Radiation Research* 22:643–661, 1964.

Reniers, B., Liu, D., Rusch, T. and Verhaegen, F. Calculation of relative biological effectiveness of a low-energy electronic brachytherapy source. *Physics in Medicine and Biology* 53:7125–7135, 2008.

Reuvers, A. P., Greenstock, C. L., Borsa, J. and Chapman, J. D. Studies on the mechanism of chemical radioprotection by dimethyl sulphoxide. *International Journal of Radiation Biology* 24:533–536, 1973.

Roberge, M., Tudan, C., Hung, S. M., Harder, K. W., Jirik, F. R. and Anderson, H. Antitumor drug fostriecin inhibits the mitotic entry checkpoint and protein phosphatase 1 and 2A. *Cancer Research* 54:6115–6121, 1994.

Roots, R. and Okada, S. Protection of DNA molecules of cultured mammalian cells from radiation-induced single-strand scissions by various alcohols and SH compounds. *Radiation Research* 21:329–342, 1972.

———. Estimation of life times and diffusion distances of radicals involved in x-ray induced DNA strand breaks or killing of mammalian cells. *Radiation Research* 64:306–320, 1975.

Ruggieri, R. and Nahum, A. E. The impact of hypofractionation on simultaneous dose-boosting to hypoxic tumor subvolumes. *Medical Physics* 33:4044–4055, 2006.

Rutkowska, E., Baker, C. and Nahum, A. A mechanistic computer simulation of normal-tissue damage in radiotherapy—Implications for dose-volume analyses. *Physics in Medicine and Biology* 55:2121–2136, 2010.

Rutkowska, E., Syndikus, I., Baker, C. and Nahum, A. E. Mechanistic modeling of radiotherapy-induced lung toxicity. *British Journal of Radiology* 85:e1242–e1248, 2012.

Sanche, L. Interaction of low-energy electrons with atomic and molecular solids. *Scanning Microscopy* 9:619–656, 1995.

———. Low energy electron damage in DNA. In *Radiation induced molecular phenomena in nucleic acids: A comprehensive theoretical and experimental analysis,* ed. M. K. Skukla and J. Leszczynski, pp. 531–575, the Netherlands: Springer Science, 2008.

Sanchez-Nieto, B. and Nahum, A. E. The delta-TCP concept: A clinically useful measure of tumor control probability. *International Journal of Radiation Oncology Biology Physics* 44:369–380, 1999.

Sanchez-Nieto, B. and Nahum, A. E. BIOPLAN: software for the biological evaluation of radiotherapy treatment plans. *Medical Dosimetry* 25:71–76, 2000.

Sanchez-Nieto, B., Nahum, A. E. and Dearnaley, D. P. Individualisation of dose prescription based on normal-tissue dose-volume and radiosensitivity data. *International Journal of Radiation Oncology Biology Physics* 49:487–499, 2001.

Saunders, M. I., Dische, S., Barrett, A., Harvey, A., Griffiths, D. and Palmar, M. Continuous hyperfractionated accelerated radiotherapy (CHART) vs. conventional radiotherapy in non-small-cell lung cancer: Mature data from the randomized multicentre trial. *Radiotherapy and Oncology* 52:137–148, 1999.

Setlow, R. B. and Pollard, E. C. *Molecular biophysics.* Reading, MA: Addison–Wesley Publ. Co., 1962.

Seymour C. B. and Mothersill, C. Radiation-induced bystander effects—Implications for cancer. *Nature Reviews Cancer* 4:158–164, 2004.

Sharma, A., Singh, K. and Almasan, A., Histone H2AX phosphorylation: A marker for DNA damage. *Methods in Molecular Biology* 920:613–626, 2012.

Short, S., Mayes, C., Woodcock, M., Johns, H. and Joiner, M. C. Low dose hypersensitivity in the T98G human glioblastoma cell line. *International Journal of Radiation Biology* 75:847–855, 1999.

Sinclair, W. K. and Morton, R. A. Variations in x-ray response during the division cycle of partially synchronized Chinese hamster cells in culture. *Nature* 199:1158–1160, 1963.

Skarsgard, L. D., Wilson, D. J. and Durand, R. E. Survival at low dose of asynchronous and partially synchronized Chinese hamster V79-171 cells. *Radiation Research* 133:102–107, 1993.

Smith, L. G., Miller, R.C., Richards, M., Brenner D. J. and Hall, E. J. Investigation of hypersensitivity to fractionated low-dose radiation exposure. *International Journal of Radiation Oncology Biology Physics* 45:187–191, 1999.

Steel, G. G. *Basic clinical radiobiology.* London: Arnold, 2002.

———. Dose fractionation in radiotherapy, ch. 9, in *Handbook of radiotherapy physics—Theory and practice,* ed. P. Mayles, A. Nahum and J-C. Rosenwald. New York, London: Taylor and Francis, pp. 163–177, 2007a.

———. Radiobiology of normal tissues, ch. 8, in *Handbook of radiotherapy physics—Theory and practice,* ed. P. Mayles, A. Nahum and J-C. Rosenwald. New York, London: Taylor and Francis, pp. 149–162, 2007b.

Steel, G. G., Deacon, J. M., Duchesne, G. M., Horwich, A., Kelland, L. R. and Peacock, J. H. The dose-rate effect in human tumor cells. *Radiotherapy and Oncology,* 9:299–310, 1987.

Steel, G. G. and Peacock, J. H. Why some human tumors are more radiosensitive than others. *Radiotherapy and Oncology* 15:63–72, 1989.

Stewart, F. A. and Dörr, W. Milestones in normal tissue radiation biology over the past 50 years: From clonogenic cell survival to cytokine networks and back to stem cell recovery. *International Journal of Radiation Biology* 85:574–586, 2009.

Stobbe, C. C., Park, S. J. and Chapman, J. D. The radiation hypersensitivity of cells at mitosis. *International Journal of Radiation Biology* 78:1149–1157, 2002.

Stock, R. G., Stone, N. N., Tabert, A., Iannuzzi, C. and DeWyngaert, J. K. A dose-response study for I-125 prostate implants. *International Journal of Radiation Oncology Biology Physics* 41:101–108, 1998.

Stone, R. S. Clinical experience with fast neutron therapy. *American Journal of Roentgenology* 59:771–785, 1984.

Sutherland, R. M. Cell and environmental interactions in tumor microregions: The multicell spheroid model. *Science* 240:177–184, 1988.

Takata, H., Hanafusa, T., Mori, T., Shimura, M., Iida, Y., Ishikawa, K., Yoshikawa, K., Yoshikawa, Y. and Maeshima, K. Chromatin compaction protects genomic DNA from radiation damage. *PLoS ONE* 8 (10): e75622.doi:10.1371/journal.pinc.oo75622, 2013.

Terasima, T. and Tolmach, L. J. Changes in x-ray sensitivity of HeLa cells during the division cycle. *Nature* 190:1210–1211, 1961.

———. Variations in several responses of HeLa cells to x-radiation during the division cycle. *Biophysical Journal* 3:11–33, 1963.

Thames, H. D. and Hendry, J. H. *Fractionation in radiotherapy.* New York: Taylor & Francis, 1987.

Thames, H. D., Jr., Withers, H. R., Peters, L. J. and Fletcher, G. H. Changes in early and late radiation responses with altered dose fractionation: Implications for dose-survival relationships. *International Journal of Radiation Oncology Biology Physics* 8:219–226, 1982.

Thiagarajan, A., Caria, N., Schöder, H., Iyer, N. G., Wolden, S., Wong, R. J., Sherman, E., Fury, M. G. and Lee, N. Target volume delineation in oropharyngeal cancer: Impact of PET, MRI and physical examination. *International Journal of Radiation Oncology Biology Physics* 83:220–227, 2012.

Thorndyke, C., Meeker, B. E., Thomas, G., Lakey, W. H., McPhee, M. S. and Chapman, J. D. The radiation sensitivities of R3327-H and R3327-AT rat prostate adenocarcinomas. *Journal of Urology* 134:191–198, 1985.

Till, J. E. and McCulloch, E. A. A direct measurement of the radiation sensitivity of normal mouse bone marrow cells. *Radiation Research* 14:213–222, 1961.

Tobey, R. A. Production and characterization of mammalian cells reversibly arrested in G_1 by growth in isoleucine-deficient medium. *Methods in Cell Biology* 6:67–112, 1973.

Tobey, R. A., Valdez, J. G. and Crissman, H. A. Synchronization of human diploid fibroblasts at multiple stages of the cell cycle. *Experimental Cell Research* 179:400–416, 1988.

Tobias, C. A., Blakely, E. A., Chang, P. Y., Lommel, L. and Roots, R. Response of sensitive human ataxia and resistant T-1 cell lines to accelerated heavy ions. *British Journal of Cancer* 49 (Suppl. VI):175–185, 1984.

Tobias, C. A., Blakely, E. A. Ngo, F. Q. H. and Chatterjee, A. Repair-misrepair (RMR) model for the effect of single and fractionated doses of heavy accelerated ions. *Radiation Research* 74:589, 1978 (abstract from 26th annual meeting of RRS).

Tobleman, W. T. and Cole, A. Repair of sublethal damage and oxygen enhancement ratio for low-voltage electron beam irradiation. *Radiation Research* 60:355–360, 1974.

Todd, P., Heavy-ion irradiation of cultured human cells. *Radiation Research* 7 (Suppl.):196–207, 1967.

———. Radiobiology with heavy charged particles directed at radiotherapy. *European Journal of Cancer* 10:207–210, 1974.

Tolmach, L. J. Growth patterns of x-irradiated HeLa cells. *Annals of New York Academy of Sciences* 95:743–757, 1961.

Urtasun, R. C., Palmer, M., Kinney, B., Belch, A., Hewitt, J. and Hanson, J. Intervention with the hypoxic tumor cell sensitizer etanidazole in the combined modality treatment of limited stage small-cell lung cancer: A one-institution study. *International Journal of Radiation Oncology Biology Physics* 40:337–342, 1998.

Uzan, J. and Nahum, A. E. Radiobiologically guided optimisation of the prescription dose and fractionation scheme in radiotherapy using BioSuite. *British Journal of Radiology* 85:1279–1286, 2012.

Valdagni, R., Italia, C., Montanaro, P., Lanceni, A., Lattuada, P., Magnani, T., Fiorino, C. and Nahum, A. Is the alpha-beta ratio of prostate cancer really low? A prospective, non-randomized trial comparing standard and hyper-fractionated conformal radiation therapy. *Radiotherapy and Oncology* 75:74–82, 2005.

Vos, O. and Kaalen, C. A. C. Protection of tissue-culture cells against ionizing radiation. II. The activity of hypoxia, dimethyl sulfoxide, dimethyl sulfone, glycerol and cysteamine at room temperature and at −196°C. *International Journal of Radiation Biology* 5:609–621, 1962.

Walicka, M. A., Adelstein, S. J. and Kassis, A. I. Indirect mechanisms contribute to biological effects produced by decay of DNA-incorporated iodine-125 in mammalian cells *in vitro*: Clonogenic survival. *Radiation Research* 149:142–146, 1998a.

———. Indirect mechanisms contribute to biological effects produced by decay of DNA-incorporated iodine-125 in mammalian cells *in vitro*: Double-strand breaks. *Radiation Research* 149:134–141, 1998b.

Ward, J. F. Molecular mechanisms of radiation-induced damage to nucleic acids. *Advances in Radiation Biology* 5:181–239, 1975.

———. Some biochemical consequences of the spatial distribution of ionizing radiation produced free radicals. *Radiation Research* 86:185–195, 1982.

———. DNA damages produced by ionizing radiation in mammalian cells: Identities, mechanisms of formation and reparability. *Progress in Nucleic Acid Research and Molecular Biology* 35:95–125, 1988.

Wasserman, T. H. and Chapman, J. D. Radiation response modulation. Part A. Chemical sensitizers and protectors. In *Principles and practice of radiation oncology*, ed. C. A. Perez, L. W. Brady, E. C. Halperin, and R. K. Schmidt-Ullrich, pp. 663–698. Philadelphia: Lippincott Williams & Wilkins, 2004.

Webb, S. Conformal and intensity-modulated radiotherapy, ch. 43, in *Handbook of radiotherapy physics—Theory and practice,* ed. P. Mayles, A. Nahum and J-C. Rosenwald, pp. 943–985, New York, London: Taylor & Francis, 2007.

Webb, S. and Nahum, A. E. A model for calculating tumor control probability in radiotherapy including the effects of inhomogeneous distributions of dose and clongenic cell density. *Physics in Medicine and Biology* 38:653–666, 1993.

Weisenthal, L. M. and Lippman, M. E. Clonogenic and nonclonogenic *in vitro* chemosensitivity assays. *Cancer Treatment Reports* 69:615–632, 1985.

West, C. M. L., Davidson, S. E., Roberts, S. A. and Hunter, R. D. Intrinsic radio-sensitivity and prediction of patient response to radiotherapy for carcinoma of the cervix. *British Journal of Cancer* 68:819–823, 1993.

———. The independence of intrinsic radiosensitivity as a prognostic factor for patient response to radiotherapy of carcinoma of the cervix. *British Journal of Cancer* 78:550–553, 1997.

Whitmore, G. F. and Gulyas, S. Synchronization of mammalian cells with tritiated thymidine. *Science* 151:691–694, 1966.

Withers, H. R., The four R's of radiotherapy. *Advances in Radiation Biology* 5:241–271, 1975.

Withers, H. R., Thames, H. D., Jr. and Peters, L. J. A new isoeffect curve for change in dose per fraction. *Radiotherapy and Oncology* 1:187–191, 1983.

Withers, H. R., Taylor, J. M. and Maciejewski, B., Treatment volume and tissue tolerance. *International Journal of Radiation Oncology Biology Physics* 14:751–759, 1988.

Wolffe, A. *Chromatin structure and function.* San Diego, CA: Academic Press, 1998.

Wouters, B. G., Sy, A. M. and Skarsgard, L. D. Low-dose hypersensitivity and increased radioresistance in a panel of human tumor cell lines with different radiosensitivity. *Radiation Research* 146:399–413, 1996.

Xu, B. and Kastan, M. G. Analyzing cell cycle check points after ionizing radiation. *Methods in Molecular Biology* 281:283–292, 2004.

Yamashita, K., Yasuda, H., Pines, J., Yasumoto, K., Nishitani, H., Ohtsubo, M., Hunter, T., Sugimura, T. and Nishimoto, T. Okadaic acid, a potent inhibitor of type I type 2A protein phosphatases, activates cdc2/H1 kinase and transiently induces a premature mitosis-like state in B\hk21 cells. *EMBO Journal* 9:4331–4338, 1990.

Yaromina, A., Krause, M., Thames, H., Rosner, A., Krause, M., Hessel, F., Grenman, R., Zips, D. and Baumann, M. Pre-treatment number of clonogenic cells and their radiosensitivity are major determinants of local tumor control after fractionated radiation. *Radiotherapy and Oncology* 83:304–310, 2007.

Yasui, L. S., Hughes, A. and DeSombre, E. R. Production of clustered DNA damage by I-125 decays. *Acta Oncologica* 39:358–366, 2008.

Yeh, K. A., Biade, S., Lanciano, R. M., Brown, D. Q., Fenning, M. C., Babb, J. S., Hanks, G. E. and Chapman, J. D. Polarographic needle electrode measurements of oxygen in rat prostate carcinomas: Accuracy and reproducibility. *International Journal of Radiation Oncology Biology Physics* 33:111–118, 1995.

Zaider, M. and Hanin, L. Tumor control probability in radiation treatment. *Medical Physics* 38:574–583, 2011.

Zellmer, D. L., Chapman, J. D., Stobbe, C. C. and Das, I. J. Radiation fields backscattered from material interfaces: I. Biological effectiveness. *Radiation Research* 150:406–415, 1998.

Index